MW00582297

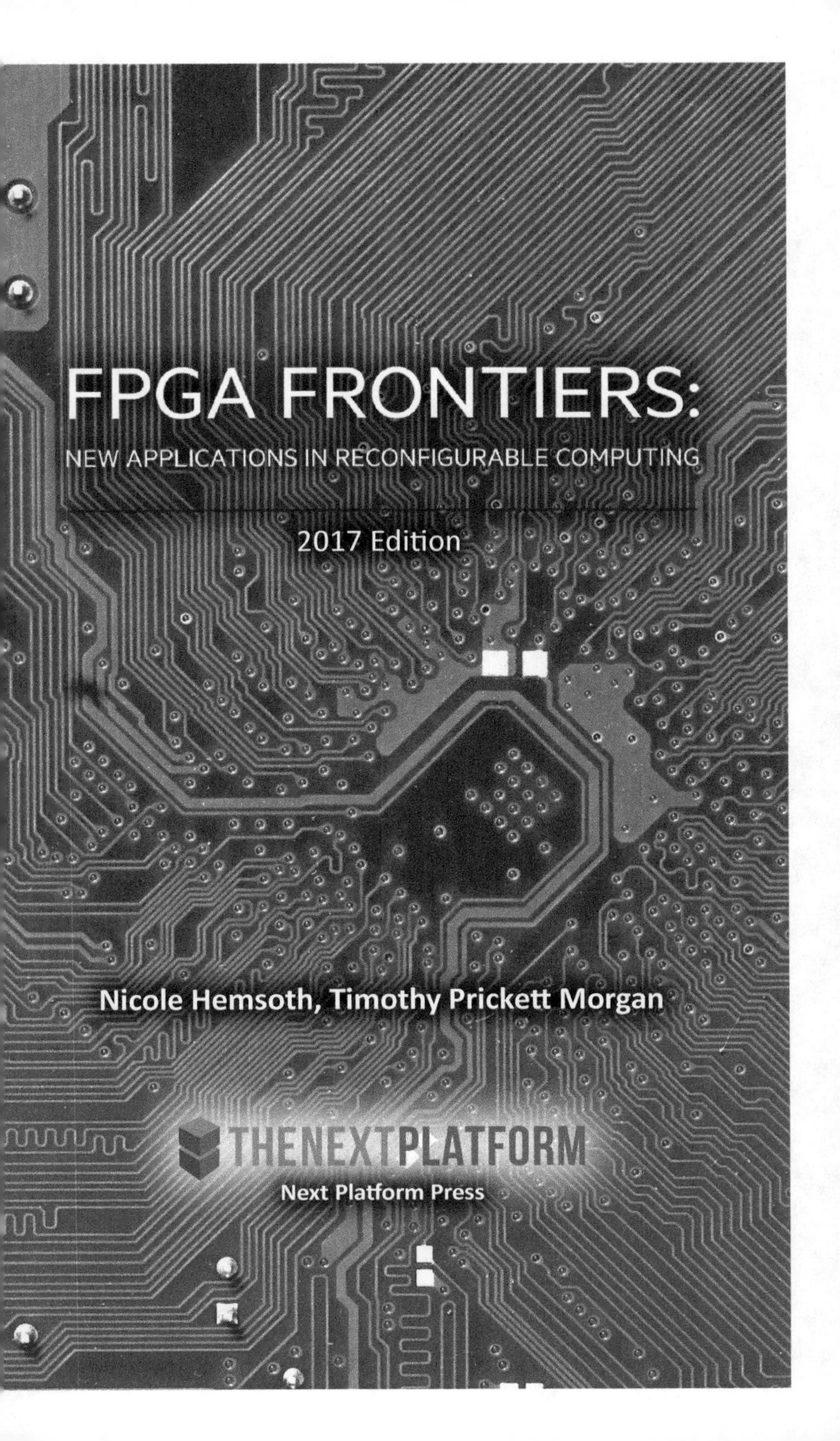
FPGA FRONTIERS:
NEW APPLICATIONS IN RECONFIGURABLE COMPUTING
2017 Edition
Nicole Hemsoth, Timothy Prickett Morgan
THENEXTPLATFORM
Next Platform Press

FPGA Frontiers: New Applications in Reconfigurable Computing, 2017 Edition

Book Design by Pamela Trush, Delaney-Designs.com

Cover Design by Fenke Fros

First Print Edition, January, 2017

ISBN: 978-0-692-83546-3

Acknowledgments

First and foremost, we would like to thank FPGA maker, Xilinx, for its initial support in the creation of the book, which is rooted in a year of coverage across the entire FPGA ecosystem. Like other vendors and researchers in this exciting space, Xilinx has helped us understand broader market and technology trends to help us anticipate what lies ahead in 2017 and beyond. Xilinx also allowed readers to access the full book in download format for the first two weeks prior to publication.

We would also like to thank our partners at The Register *for the ongoing support of our efforts on* The Next Platform *and in our foray into print publishing for select topics.*

Our readers deserve a special thanks as well since it is their input and feedback that drives our analysis of the technology market at the high end.

Table of Contents

Guest Sponsor Introduction

Ivo Bolsens, Chief Technology Officer, Xilinx

There is little doubt that this is a new era for FPGAs.

While it is not news that FPGAs have been deployed in many different environments, particularly on the storage and networking side, there are fresh use cases emerging in part due to much larger datacenter trends. Energy efficiency, scalability, and the ability to handle vast volumes of streaming data are more important now than ever before. At a time when traditional CPUs are facing a future where Moore's Law is less certain and other accelerators and custom ASICs are potential solutions with their own sets of expenses and hurdles, FPGAs are getting a serious second look for an ever-growing range of workloads.

FPGAs have always been a multi-market solution, but the market is more varied in terms of compute requirements, yet more homogenous than ever in terms of those basic requirements for efficiency, speed, scalability, and usability. A great deal of work has gone into improving the profile along every single one of these trajectories, not the least of which is programmability—the assumed Achilles' heel for reprogrammable devices, but to far less of a degree than ever in the history of FPGAs.

In his fifteen years as FPGA maker Xilinx's Chief Technology Officer, Ivo Bolsens has watched as reprogrammable devices have moved from being the glue logic to the heart of full systems. "This is because it is now possible to incorporate so many rich capabilities into the devices and governing software framework. In terms of the future, we think we have also made it clear that the broader compute world has an opportunity with FPGAs as well given larger trends in the datacenter, most notably performance per watt and overall scalability. Scaling up and down at the same time as having programmability mean more in the wake of these trends, and this of course maps well to FPGAs," he explains.

The goal for companies like Xilinx is to deliver high compute density and capability in an FPGA to meet these growing workload requirements. These two areas are the real starting point, especially as we look at emerging needs in both cloud and machine learning. Many applications in both of these spheres have data flow and streaming data processing needs, and this is exactly where FPGAs shine with less power consumption than other accelerator architectures or CPU only approaches. "This is because the data and compute are side by side without heavy, expensive data movement between memories—a feature that is most important for machine learning," Bolsens notes.

These same aspects that make FPGAs an attractive fit for the emerging workloads map well to other areas where reprogrammable logic has a major role. Network processing, security, deep packet inspection are all important areas for FPGAs. Also on the horizon are even larger trends feeding more work to the FPGA in terms of adding greater levels of intelligence into both the network and storage layers. The opportunities here are huge; the ability to bring the compute closer to the storage and data for doing networking functions is a game-changing capability that CPUs cannot touch performance and efficiency-wise. The emerging trend toward network function virtualization alone represents a major opportunity for FPGAs in tandem with CPUs and although it is a different use case from machine learning, video transcoding, and other emerging workloads, it shows how and why the FPGA can hum against streaming data in a way other accelerators or CPU-only approaches cannot.

The historical challenge for FPGAs entering into new and emerging markets has been the programming environment, but this problem is being solved by major step-changes. "Much of our research and development organization has been focused on the future of making reprogrammable devices programmable, and with OpenCL and critical insights we have had over the years to make these more approachable, we are now moving toward general purpose (in terms of usability) devices. To put this into some perspective, it is useful to understand where Xilinx began with FPGAs and where we are now for both the hardware and software end users of our devices," says Bolsens.

The challenge ten years ago was to bridge the gap between the hardware and software sides of the development house. "We wanted to make sure it was possible to unleash the full potential of the hardware platform without exposing the software people to all of the gritty details of the underlying hardware, beginning with an effort to move from Verilog and HDL to higher level abstractions," Bolsens recalls. "A decade ago, we actually had to sell this concept inside of Xilinx—this idea that we could build hardware

with fewer lines of code and with a high-level synthesis approach that allowed us to see functions and map them into hardware. Ten years ago, this was not a need. Today it is, and we have responded."

So much has changed in terms of FPGA usability in the last decade that it is quite stunning to stop and see that big picture. If you are a software developer these days in a datacenter environment and you are writing your C++ programs to run in the cloud with FPGA acceleration it is now seamless. It is now no longer a major hurdle to run in a heterogeneous platform with your code running on the CPU, GPU, FPGA, or all of these together. You can now get the benefits of the FPGA without having to deal with the requirements of specialized knowledge about programming those FPGAs. This is a long way to have come for these disparate hardware and software people—from those beginnings to this more seamless environment.

Of course, there is still a great deal of work to do. One of the things that is still missing for FPGAs compared to other accelerators—and something companies like Xilinx understand they still need to address for some of the emerging workloads like machine learning and deep learning—is providing the libraries needed. "Today, people do not write software from scratch; they are using many libraries and compared to other platforms, this is where catching up needs to be done. The goal is to leverage our ecosystem and partners, and of course, leverage our in-house software expertise to build around this gap and ensure new application areas can quickly onboard with FPGA acceleration," Bolsens says.

Even though we talk about programmability and the availability of key libraries, the biggest hurdle for FPGAs is also their most attractive point; they allow an immense amount of freedom. For those who are skilled in navigating FPGA use this is the benefit, along with performance per watt—this flexibility. However, that tremendous amount of programmability means the user experience can be more complex. As Bolsen explains, "There are many ways to mess up but we are addressing this with additional investments in templates to make developers as efficient and productive as possible in key domains so they can adhere to best practices, avoid mistakes, and get around otherwise longer efforts riding a learning curve."

What sets FPGAs apart is not only the flexibility, energy efficiency, and price-performance profile we have described, but also the fact that FPGA makers understand interconnects better than anyone else in the semiconductor world. "We understand how to build an interconnect infrastructure that is programmable, can connect any function to any other function by programming a device, and with that kind of infrastructure in place, the potential applications abound," Bolsen notes.

We look forward to witnessing where FPGAs find a place in the new world of applications that are being driven by big data and see a path to this internally. As we will highlight in the course of this book, there are numerous opportunities for emerging application areas and despite some roadblocks and challenges, there is great hope on the FPGA horizon.

Introduction

For most of the more than three decades that field programmable gate arrays have been available as compute elements, proponents of these exotic devices have been awaiting the day when all of the stars of technology and economics would align and their approach to creating logic devices would knock application-specific integrated circuits, or ASICs, out of favor.

It never happened, and we are all still waiting. But perhaps we are waiting for the wrong thing.

Not just because we like a good competitive battle between technologies and the fight for revenue and profits, but because the arguments that FPGA makers gave for why their devices should reign supreme had an elegance and a logical consistency to them. While FPGA makers originally downplayed the malleability of their devices, the idea that you can take a chip on the most advanced process nodes available and make it do anything just about any ASIC can do – that you can in essence recycle transistors and change the personality of the FPGA through something akin to, but not quite the same as, programming – is just plain appealing. And we believe it should have won over the market a long time ago, just like FPGA makers like Xilinx and Altera, also believed. The malleability of an FPGA should trump the raw circuit speeds and feeds of an ASIC.

The funny bit is that for a certain part of the computing workload in network and storage devices in the datacenter as well as consumer devices, FPGAs did carve out a slice. That is why we are able to even contemplate a larger role for FPGAs in compute within the datacenter today. Had there not been myriad smaller bits of computing that were best implemented on an FPGA because a dedicated ASIC was too expensive and because running that work as a bit of software on a generic CPU provided too low performance at too high of a cost, the FPGA market would never have coalesced at all.

Rather than kill the ASIC, the very existence of the FPGA alternative has performed a vital role in pushing workloads towards specialized ASICs, where functions are hard coded (like a switch chip), towards general purpose ASICs (like a Xeon processor) that implement functions in software, or towards a hybrid application-specific standard product (ASSP) system on chip device, which mixes fixed function ASICs with general purpose programmable ASICs like a processor. The lines are considerably more blurred when you consider that what we call FPGAs are actually an amalgamation of an FPGA plus other ASICs like a Power or ARM processor, providing both the malleability of the FPGA and the programmability of the CPU all in the same package.

To a certain way of looking at it, the hybrid world that we are contemplating in the compute portion of the datacenter has already happened at a smaller scale of system on chip devices that are used in a range of commercial and consumer machines. The reason is simple: They are power constrained and cost constrained, or both, and they need the various attributes of multiple kinds of compute devices. In some cases, a machine is mixing all four kinds of compute – CPU, FPGA, GPU, and DSP – although not on a single device. At least not yet. But the mix is necessary to provide the most efficient machine possible given power and budget constraints. What has applied to an Apple iPhone for years applies equally well to a compute farm or a datacenter, and this is not particularly surprising. Sooner or later, the Moore's Law limits that have shaped the engineering of our smartphones was going to shape the engineering of our datacenters. We can make datacenters bigger, to be sure, but we can't get dozens of megawatts of power to them. Just like we can make a smartphone bigger, but we want something that fits in our hands and that has a battery life of at least a day as it does more and more sophisticated things with each successive generation.

Because of this, and because of the deep experience that network equipment makers and storage array makers have with FPGAs already, we think the time is ripe and right for FPGAs to take their rightful place in datacenter compute. In this and successive chapters, we will endeavor to look at where FPGAs will mesh with existing and emerging applications with an emphasis on their potential role in enterprise, machine learning, and other forward-looking applications. In short, those areas that surpass the traditional role of FPGAs in networks, storage, and other gear.

CHAPTER ONE

The FPGA Market: Past, Present, and Future

The FPGA has a time to market and cost advantage over specialized ASICs and a performance advantage in most cases compared to running software atop CPUs to perform certain functions, and that is why we think they will be sprinkled around the datacenter in server, switching, and storage layers and accelerating functions all throughout the workflow.

It is important to not get overly excited, however. Particularly with Intel shelling out $16.7 billion in December 2015 to buy FPGA maker Altera. In late 2014, when Altera was a public company, the company's top brass pegged the parallel computing opportunity in the datacenter at around $1 billion for hybrid machines that mixed CPUs and FPGAs, compared to around $250 million for hybrid CPU-GPU machines and around $9 billion for processors running applications directly on CPUs.

People are familiar with CPUs, of course, and finding a C programmer is not too difficult, so there is not much risk or pain in using straight up CPUs to perform computing tasks. But the energy efficiency is relatively low, particularly for workloads that are compute intensive and inherently parallel. The CPU-GPU hybrid is less familiar to programmers but the combination is considerably more power efficient. By Altera's estimation, programming a hybrid CPU-FPGA system using OpenCL is easier for programmers than use Nvidia's CUDA environment (something that is arguable for sure), but hard coding the FPGAs in HDL is quite difficult and hence the need for OpenCL or some other abstraction layer to automate the offloading of functions from the CPU to the FPGA.

That $1 billion datacenter opportunity includes sales by Altera, Xilinx, and other FPGA suppliers, and it meshes more or less with the way Intel cased the market in June 2015 when it made the deal to buy Altera. In 2014, ahead of the Intel acquisition, 16 percent of Altera's $1.9 billion revenues came from the compute, network, and storage business that is related to datacenters, which works out to $304 million. Telecom and wireless equipment makers, who make systems that stay in the field for one or two decades and who want power efficiency, low cost, and the kind of malleability that FPGAs provide without needing an operating system or the software overhead of a CPU for those functions, comprised 44 percent Altera's sales, or $835 million. Another 22 percent of Altera's revenues, or $418 million, came from makers of industrial controls, military equipment, and automobiles, which have the same constraints and therefore choose FPGAs for some of their workloads.

At the time, Intel pegged the total addressable market for all kinds of chips at $115 billion for all of 2014, with programmable logic devices (dominated by FPGAs) accounting for about 4 percent of the pie, ASICs at 18 percent, and the remainder being the hodge podge of ASSPs. Within the programmable logic devices segment, Intel reckoned Altera had 39 percent of the $4.8 billion market, with Xilinx getting 49 percent and a handful of other vendors brining in the remaining 12 percent.

Intel did not buy Altera just because the FPGA business is growing almost as fast as its own Data Center Group, which supplies chips, chipsets, and motherboards to makers of servers, storage, and switches. Intel did so because as Moore's Law is slowing down, FPGAs are more and more of a competitive threat. Each FPGA accelerator – and indeed, any GPU or DSP accelerator – that is installed in a datacenter is not just one, but probably a couple of Xeon CPUs that are not going to be installed. Intel can't keep making Xeons bigger and bigger with more cores and more accelerators on them, and so it must have therefore concluded that it needs to have FPGA acceleration in its own right. It is better to sell $500 million in FPGAs in the datacenter today and maybe $1 billion or more of them years hence and sacrifice perhaps two or three times that in Xeon revenues than just lose the Xeon revenues to someone else.

According to Intel's forecasts, the company is projecting a fairly linear growth rate for the FPGA business between now and 2023, and we are always suspicious of such linear forecasting. But the FPGA business has grown more or less over time (it is about 2.5X larger than it was fifteen years ago) and Intel expects it to almost double between 2014 and 2023. With a compound annual growth rate of 7 percent between 2014 and 2023 as Intel is forecasting, the revenues should be just shy of $8.9 billion at the end of the forecast period. Interestingly, Intel's forecast is not projecting that the share of FPGA revenues from datacenter compute (servers, switching, and networking) is going to change by much. Take a look:

If market shares didn't change between Altera and Xilinx, and the portion of Altera's revenues remained the same for networking, compute, and storage, then this piece of the Altera business would be somewhere around $560 million in 2023. We think that such a number underestimates the pressure that datacenters will be under to provide more efficient and flexible computing, and that the prospects for FPGAs are considerably better than this forecast suggests. That said, many proponents of FPGA technology have been waiting for the day when the FPGA would get its rightful share of compute in the datacenter, and we think they will be pleasantly surprised to find it happening soon. The irony is that Intel itself, which is an expert in programming FPGAs and using hardware description languages by virtue of its role as an ASIC manufacturer, will be accelerating the adoption of FPGAs as accelerators, both as free-standing discrete compute elements and as hybrid CPU-FPGA devices.

It is difficult not to place too much emphasis on the Intel, Altera news from 2016 since it portends massive growth for FPGAs. This can do nothing other than serve the other FPGA makers, few as they are—at least in the short term.

Intel's acquisition and subsequent statements about the potential for FPGAs for a growing range of workloads is worth exploring in some detail, which we will do to set the stage for the rest of this book. That acquisition marks a defining event for the future of FPGAs as it shows Intel recognizes the vast potential. Just as other moves by major companies show, including Amazon Web Services, which is providing Xilinx FPGAs with the expectation of big market momentum, the future of FPGAs as compute accelerators is strong.

Accelerating workloads in the enterprise, boosting search inside of hyperscale datacenters, and goosing the performance of HPC simulations are all hot areas for FPGAs. The addition of machine learning and deep learning to the application ranks marks another line that FPGAs can cross for a stronger 2017. As we will detail throughout the course of this book, this is happening for a few reasons.

First, the software stacks for programming FPGAs have evolved, and Altera in particularly has been instrumental in grafting support for FPGAs onto the OpenCL development environment that was originally created to program GPUs and then was tweaked to allow it to turn GPUs into offload engines for CPUs. (Not everyone is a big fan of OpenCL. Nvidia has created its own CUDA parallel programming environment for its Tesla GPU accelerators, and also in mid-2016, SRC Computers, a company that has been delivering hybrid CPU-FPGA systems in the defense and intelligence industries since 2002, launched into the commercial market with its own Carte programming environment, which turns C and Fortran programs into the FPGA's hardware description language (HDL) automatically.

The other factor that is driving FPGA adoption is the fact that getting more performance out of a multicore CPU is getting harder and harder as the process shrink jumps get smaller for chip manufacturing techniques. Performance jumps are being made, but mostly in expanding the performance throughput of CPUs, not the individual performance of a single CPU core. (There are some hard-fought architectural enhancements, we know.) But both FPGA and GPU accelerators offer a more compelling improvement in performance per watt. Hybrid CPU-FPGA and CPU-GPU systems can offer similar performance and performance per watt, at least according to tests that Microsoft has ran, on deep learning algorithms. The GPUs run hotter, but they do roughly proportionally more work and at the system level, offer similar performance per watt.

That increase in performance per watt is why the world's most powerful supercomputers moved to parallel clusters in the late 1990s and why they tend to be hybrid machines right now, although Intel's massively parallel "Knights Landing" Xeon Phi processors are going to muscle in next to CPU-GPU hybrids. With Altera FPGA coprocessors and Knights Landing Xeon Phi processors, Intel can hold its own against the competition at the high end, which is shaping up to be the OpenPower collective that brings together IBM Power processors, Nvidia Tesla coprocessors, and Mellanox Technologies InfiniBand networking.

Intel has to believe that the workloads in the hyperscale, cloud, enterprise, and HPC markets will grow fast enough to let it still ramp its compute business, or that it has no choice but to be the seller of FPGAs or else someone else will be, or perhaps a little of both. But Intel is not talking about it this way.

"We do not consider this a defensive play or more," Intel's CEO, Brian Krzanich said in a press conference in the wake of the Altera acquisition news. "We look at this, in terms of both IoT and the datacenter, as expansive. These are products that our customers want built. We have said that 30 percent of the cloud workloads will be on these products as we exit this decade, and that is an estimate on our part on how we see trends moving and where we see the market going. This is about providing the capability to move those workloads down into the silicon, which is going to happen one way or another. We believe that it is best done with the Xeon processor-FPGA combination, which will clearly have the best performance and cost for the industry. In IoT, it is about expanding into new available markets against ASICs and ASSPs, and with datacenter moving those workloads down into silicon and enabling the growth of the cloud overall."

Krzanich explains, "You can think of an FPGA as a large sea of gates that they can program now. And so if they think that their algorithm will change over time as they learn and get smarter, or if they want to get to be more efficient and they don't have enough volume to have a single workload on a single piece of silicon then they can use an FPGA as accelerators of multiple segments – doing facial search at the same time as doing encryption – and we can basically on-the-fly reprogram this FPGA, literally within microseconds. That gives them a much lower cost and a much greater level of flexibility than a single customized part that you would need to have quite a bit of scale to need.

Intel sees a much larger opportunity than this, and CEO Brian Krzanich said when the deal was announced that up to a third of cloud service providers could be using hybrid CPU-FPGA server nodes for their workloads by 2020. This is an astounding statement, given that Altera itself pegged the FPGA opportunity in the datacenter at something around $1 billion in its own forecasts from late 2014. That's about three times the current revenue run rate for Nvidia's Tesla compute engines. Intel showed a prototype Xeon-FPGA chip that put the two devices on the same package back in early 2014, and the plan was to get it out the door by the end of 2016 with a ramp through 2017; the idea was to get a Xeon with FPGA circuits on the die "shortly after that," as Data Center Group general manager Diane Bryant put it at the time. On the call announcing the Altera deal, Krzanich did not say anything about the timing of this Xeon-FPGA device, but did say that Intel would create a hybrid Atom-FPGA device aimed at the Internet of Things market that would be a monolithic die; Intel is examining if it needs to do a single-package hybrid in the interim based on Atoms and Altera FPGAs.

Not surprisingly, FPGAs were the hot topic of conversation in early 2016 when Jason Waxman, general manager of Intel's Cloud Infrastructure Group, participated in a conference call to talk about Intel's datacenter business with the research analysts at Pacific Crest Securities. First off, Waxman confirmed that Intel is already sampling that hybrid Xeon-FPGA compute engine to key cloud service providers, although he did not name names or give out any of the specs on the device.

Importantly, Waxman spoke at length and clarified what is driving Intel to acquire Altera and jump into programmable computing devices. And Intel clearly wants to make FPGAs more mainstream, even if that might cannibalize some of its Xeon business in the datacenter. (We think because Intel believes that such cannibalization is inevitable, and the best way to control it is to make FPGAs part of the Xeon lineup.)

"I think there are a number of things that can go into the acquisition, and a number of them are beyond the Data Center Group," Waxman said. "One is that there is an underlying core business that tends to be driven by manufacturing lead advantage, and we seem to have a pretty good handle on that, so there seems to be some good synergy there. There is also the Internet of Things Group that has a strong interest as well. But for us, one of the things that we started to see is that with the expansion of workloads that are done at massive scale – something like machine learning, certain network functions – there is increasingly interest in seeing how you, if you are doing it at scale, get a higher degree of performance. So we are on the cusp of realizing that if we can get some breakthroughs in performance, we can potentially take an FPGA from something that is a niche today in datacenter applications to something that is much more broad, and we see this as a great opportunity. In the Data Center Group, the synergy we see is taking the FPGA and making it a companion to the CPU and expanding our silicon footprint and being able to solve problems for cloud service providers and other types of large-scale applications."

"We see a path to accelerating machine learning, to accelerate storage encryption, to accelerate network functions. We know because we are very deep into those workloads and we now see the opportunity to do it. Now, FPGAs have traditionally been kind of difficult, limited to the far-out expertise, because you are writing RTL. We are a company that writes RTL all the time, so we can solve that problem."

The key applications where Intel thinks there is initial and presumably large demand for FPGA acceleration include machine learning, search engine indexing, encryption, and data compression. These tend to be very targeted use cases, not general purpose ones, as Waxman pointed out. These are the workloads that Krzanich was no doubt referring to when he said that a third of cloud service providers would be using FPGA acceleration within five years.

Everyone has been lamenting how difficult it is to program FPGAs, but Intel is not daunted by this, and without revealing too much about Intel's plans, he did offer some insight into why and what possible actions it might take to make FPGAs more accessible.

"I think the thing that we have that is unique, that other people would not be able to go deliver, is the ability to understand those workloads and to be able to drive acceleration," said Waxman."We see a path to accelerating machine learning, to accelerate storage encryption, to accelerate network functions. We know because we are very deep into those workloads and we now see the opportunity to do it. Now, FPGAs have traditionally been kind of difficult, limited to the far-out expertise, because you are writing RTL. We are a company that writes RTL all the time, so we can solve that problem. We can make it performant and we can lower that barrier to entry. The third piece is really the volume economics, and that is all about integration and manufacturing prowess. So we look at the barriers that have kept it a niche, and we have a path to overcome those barriers. We have some interesting plans and if things go well, we can talk about those at another time."

For those of us who think that Intel is conceptualizing this as FPGAs replacing Xeons, Waxman put the kibosh on that idea entirely.

Any algorithms that need to be done repetitively and at high rates, are a natural for FPGAs, said Waxman,

"The workload or application is going to run on a CPU, and there will be an algorithm or a piece of it that will go on the FPGA," he said. "So you are not going to run the entire application on an FPGA, and that is one of the things that I think sometimes people are wondering about. Is this thing a replacement for a Xeon CPU? It is really not, it is a companion to the CPU. Take image recognition. How does a computer identify the picture of a cat on Facebook – that's the funny example. There is a lot of compute that goes behind that, but the actual application for machine learning runs on a Xeon CPU but there are certain algorithms that you are going to want to offload to an FPGA."

and we would add that any data manipulation or transformation that needs to be done at extreme low latency or on the wire is also a candidate.

Considering that Altera already makes system-on-chips that incorporate ARM processors and FPGAs, it is natural to think that Intel might be tempted to global replace ARM cores with X86 cores and do similar devices. But it doesn't look like this will happen. First, on a call going over Intel's financial results for the second quarter of 2016, Krzanich said that Intel was committed to supporting and enhancing these ARM-FPGA hybrids for Altera's existing customers.

"I think the way that we view it is that we would actually be integrating some form of FPGA into a Xeon," Waxman clarified even further. "We have talked publicly about doing a first generation in one package, but the way we will look at it going forward, depending on how things progress, would be on the same die. So we will be looking at what is the right combination based on the customer feedback. And by the way, I would still expect to see there will be some systems where there won't be integration, they will still do a system-level companion. We are not going to integrate every possible combination of Xeon with FPGAs. That would be prohibitive and we will find the right targets and balance in the market."

While Altera's toolset makes use of the OpenCL programming model to get application code converted down to RTL, the native language of the FPGA, interestingly Intel does not think that the future success of FPGAs in the datacenter is predicated on improvements in OpenCL integration with RTL tools or more widespread adoption of OpenCL.

"It is not predicated on OpenCL," Waxman said emphatically. "We do see OpenCL as a potential avenue that further broadens the applicability of FPGAs, but right now initial cloud deployments of FPGAs will probably be done by the more capable companies and none of them are asking us for OpenCL."

Intel has plans to make it easier to program FPGAs, but Waxman was not at liberty to talk about them. He did hint, however, that what Intel could do is make an RTL library available to programmers so they could call routines deployed on FPGAs, pushing it down to form the gates that implement the application routines on the gates, rather than have them create those routines by themselves. This makes a certain amount of sense, and this is precisely what Convey, which is part of Micron Technology now, did with its FPGA-accelerated systems a few years ago.

"I think there is a continuum of acceleration," Waxman says. "And what happens is, in the beginning, you may not know exactly what you are trying to accelerate and you are experimenting a little bit, and in that phase of acceleration, you want something that is a little more general purpose. As you start to really home in on what you are trying to accelerate, you are going to want something that is more efficient, that has lower power and takes less space, and that is when you are going to move into an FPGA."

Waxman then cited the work that Microsoft has done with FPGA acceleration on its "Catapult" system, which takes its Open Cloud Server and adds FPGA mezzanine cards as accelerators. We went over this research back in March, which shows how an FPGA device at 25 watts delivers better performance/watt than a set of servers using Nvidia Tesla K20 GPU accelerators at 235 watts that were tested by Google running the same image recognition training algorithms.

As we have pointed out, we have no doubts about the performance numbers that Microsoft and Google posted, but applying performance to the discrete GPU or FPGA and gauging that against its own thermal profile is not fair. You have to look at this at the server node level, and if you do that, the FPGA-assisted Microsoft server at the system level is only moderately ahead of the servers Google tested using Tesla K20s. (Those were our estimates, based on images processed per second per watt.) And this Microsoft comparison does not take cost into account, and it should. What can be honestly said is that Microsoft's Open Cloud Server does not have the juice or the cooling to use full-on Tesla GPUs. A real bakeoff would use GPU mezzanine cards somehow and include thermals, performance, and price.

But Waxman's larger point in the discussion remains the same.

"At some point, you are really going to want that thing to scream, and you are going to want to do that in a much lower power envelope. That is what we are banking on – that more optimized approach is where an FPGA is going to pay off."

The last thing to consider is that cloud business at Intel. These customers now represent about 25 percent of Data Center Group's revenue and in the aggregate their purchases are growing at about 25 percent per year. The overall Data Center Group business is projected to grow at 15 percent in 2016 and into the next couple of years. Let's do some math. Intel should post $16.6 billion in revenues in the Data Center Group in 2016 if its plan works out. That's around $4.1 billion for the cloud service providers (which includes cloud builders and hyperscalers using our language here at *The Next Platform*), and around $12.5 billion for the rest of Intel's datacenter sales. So outside of the cloud, Intel's business is growing at about 12 percent, or half the cloud rate. Intel needs to feed that cloud growth any way it can, and apparently FPGA capacity, even if it does cannibalize Xeon capacity a bit, is a better option for Intel than having GPU acceleration continue to grow as it has.

On the programming side, which is arguably one of the sticking points for wider FPGA adoption (unlike other accelerators with rich development ecosystems like CUDA for Nvidia GPUs), there is momentum to extend the ability for programmers to design at the C language level or using OpenCL versus the low-level models that plagued FPGA development in the past. But even with so many progressive points to mark wider adoption, FPGAs are still stranded just outside of mainstream adoption. We will explore this more thoroughly in our chapter on the programming problems and opportunities, but for now, this remains a sticking point. While we have talked to many of the vendors in this relatively small ecosystem, including Altera and Xilinx (the two major suppliers) about what still remains, according to long-time FPGA researcher, Russell Tessier, the glory days for FPGAs in terms of how they will hit the wider market are still ahead—and new developments across the board will mean broader adoption.

In his twenty years working with FPGAs, which he still does at the University of Massachusetts (he also had a stint at Altera and was founder of Virtual Machine Works, which Mentor Graphics acquired), he says there has been a slow transition from science project to enterprise reality for FPGAs. "A lot of this is because

Until relatively recently, FPGAs had a reputation problem. And it was not exactly inaccurate, since for any large-scale implementations, they did require specialists to program and configure, which made them off-limits in many industries, even though financial services, oil and gas, and a few other segments found the costs necessary to gain speed and throughput for specific applications.

of key improvements in design tools, with designers better able to specify their designs at a high level in addition to having the vendor tools that can be better mapped to chips." He adds that from a sheer volume perspective, the amount of logic inside the device means users are able to implement increasingly large functions, making them more attractive to a wider base.

There have been trends over last few years that are making these devices a bit easier, programmatically speaking, says Tessier. Xilinx currently encourages design at the C language level using its Vivado product. Altera also has an OpenCL environment it has developed. The key, Tessier says, is that "both companies are trying to create an environment where users can program in more familiar procedural languages like C and OpenCL rather than having to be RTL design experts in Verilog or VHDL—that's a process that's still progressing although there does seem to be more footing in the last few years that will help move things more into the mainstream."

One of the real enabling factors for FPGAs will be solving some of the memory and data movement limitations of FPGAs by meshing them with the chip and a hyper-fast interconnect. This capability might be one of the reasons some were certain the Altera and Intel talks would lead to a big purchase—the market would expand dramatically at that point and if large companies like Intel and IBM were properly motivated to push the software ecosystem for FPGAs further, the mainstreaming of the FPGA (which will likely not be as significant as GPUs, at least not yet) might happen sooner.

"Increased integration with standard core processors is certainly a key here," Tessier explains. "The barriers in the past have been languages and tools and as those get better there will be new opportunities and work with the chip vendors does open some new doors." Tessier says that because of these and other "mainstreaming" trends, the application areas for FPGAs will continue to grow, in large part because the way they're being used is changing. For instance, while financial services shops were among the first to use FPGAs for doing financial trends and stock selection analysis, the use cases are expanding as that segment now has bigger devices that can solve larger problems—and can now be strung together in larger quantities. Aside from that, other new areas for FPGAs, including in DNA sequencing, security, and encryption, and some key machine learning tasks will likely find new uses for FPGAs.

Of course, we do expect FPGAs will start big—and find their way into some of the world's largest cloud and hyperscale datacenters, a sentiment that Hamant Dhulla, VP of the datacenter division at Xilnix strongly agrees with. "Heterogeneous computing is no longer a trend, it's a reality," he told *The Next Platform* early in 2016 in the wake of the highly publicized Microsoft Catapult use cases for FPGAs (more on that later) and Intel's acquisition of Altera and subsequent statements about where they see the future of FPGAs in the datacenter.

From machine learning, high performance computing, data analytics, and beyond, there is a new day dawning for FPGAs in a more diverse range of application areas. Highly parallel workloads that ca be contained in a small power envelope and take advantage of the growing amount of on-chip memory available on FPGAs are all in the sights of FPGA makers and potential end users. Dhulla says the market potential is large enough to upset the way Xilinx has looked at its own business. For several years, storage and networking dominated the FPGA user base, but within five years, the demand on the compute side will far outpace storage and networking—both of which are expected to continue along a steady growth line.

FPGAs are set to become a companion technology in some hyperscale datacenters, Dhulla says. "We are seeing that these datacenters are separated into 'pods" or multiple racks of servers for specific workloads. For instance, some have pods set aside to do things like image resizing, as an example. Here they are finding a fit for FPGAs to accelerate this part of the workload." In other hot areas for FPGAs, including machine learning, Dhulla says that they are operating as a "cooperating" accelerator with GPUs. "There is no doubt that for the training portion of many machine learning workloads GPUs are dominant. There is a lot of compute power needed here, just as with HPC, where the power envelope tradeoff is worth it for what is required." But he says that these customers are buying tens or hundreds of GPUs instead of hundreds of thousands—those large accelerator numbers are being used on the inference part of the machine learning pipeline—and that's where the volume is. As we noted already, Nvidia is countering this with two separate GPUs (the M40 for training, the low-power M4 that plugs into pared down hyperscale servers) to counter this, but Dhulla believes FPGAs can still wick the power consumption lower and, by taking a PCIe approach, can snap into hyperscale datacenters as well.

Their SDAccel programming environment is making this more practical by offering a high-level interface to C, C++ and OpenCL, but the real path to pushing hyperscale and HPC adoption is through end user examples.

When it comes to these early users that will set the stage for the next generation of FPGA use, Dhulla points to companies like Edico Genome, which we will discuss later in this book, and how they have shifted their thinking toward FPGAs for performance and efficiency reasons, aided in part by the increasing ease of adoption. Xilinx is also currently working with production customers in other areas, including historical compute-side workloads in oil and gas and finance. But the first inkling of where their compute acceleration business is going can be seen with early customers using Xilinx FPGAs in machine learning, image recognition and analysis, and security. For instance, on the deep packet inspection front, the FPGA sits in front and all the traffic goes through it, meaning it's possible to look at each individual packet—a capability that will have implications in software-defined networking as well.

The real opportunity for FPGAs at scale lies in the cloud. Despite the clinging drawbacks of poor double-precision performance, overall price, and what is arguably a much trickier (albeit more exact) programming approach, FPGAs might have something going for them that discrete GPUs do not—a power profile that is primed for the cloud. If the FPGA vendors can convince end users that, for key workloads, their accelerators can offer a considerable performance boost (which they do in some cases), offer a programming environment that is on par complexity-wise with other accelerators (CUDA, for instance) by encouraging OpenCL development, and wick away the price concerns by offering FPGAs in the cloud, there could be hope on the horizon.

Of course, that hope is bolstered by the urge to snap FPGAs into ultra-dense servers inside cloud infrastructure versus selling on accelerated bare metal. This already happened for FPGAs in financial services, which is where they are a prime fit with their solid integer performance. But if they haven't seen mass adoption elsewhere, it's time to look to a new delivery box–not to mention some new application goodies to put inside.

Just as their GPU accelerator cousins are rallying around deep learning to make the leap to a broader set of users, particularly in the web-scale and cloud space, the FPGA set is seeing a real chance to invade the marketplace by tackling neural network and deep learning workloads. These new hosts of applications mean new markets and with the cloud removing some of the management overhead, it could mean broader adoption. Efforts to move this along are working in some key machine learning, neural network, and search applications. FPGAs are becoming more commonplace outside of the hyperscale context in areas like natural language processing (useful for a growing array of use cases from clinical settings to consumer services), medical imaging, deep packet inspection, and beyond.

Over the last year there have been a few highly publicized use cases highlighting the role of FPGAs for specific workloads, particularly in the deep learning and neural network spaces, as well as image recognition and natural language processing. For instance, Microsoft used FPGAs to give its Bing search service a 2X boost across 1,632 nodes and employed a creative 2D torus, high throughput network to support Altera FPGA-driven work. China's search engine giant, Baidu, which is also a heavy user of GPUs for many of its deep learning and neural network tasks, is using FPGAs for the storage controller on a 2,000 petabyte array that ingests between 100 terabytes to a petabyte per day. These and other prominent cases of large-scale datacenters using FPGAs, especially when they do so over GPUs, are bringing more attention to the single-precision floating point performance per watt that FPGAs bring to the table.

While some use cases, including the Baidu example, featured GPUs as the compute accelerator and FPGAs on the storage end, Altera, Xilnix, Nallatech, and researchers from IBM on the OpenPower front were showcasing where FPGAs will shine for deep learning in the cloud. The takeaway from these use cases is that the speedups for key applications were hosted inside ultra-dense machines that would melt the Xeon if a GPU was placed in concert. For tight-packed systems, they are a viable choice on the thermal front and even though there might not be as many algorithms where FPGAs can show off (compared to GPUs) this could be the beginning of a golden era for the coprocessors, especially now that there are CAPI and QPI hooks for taking advantage of shared memory on FPGA-boosted systems.

If you ask Altera, Xilinx, and others, this is happening because of what we can call the "three P's of FPGA adoption" – performance, power, and price. In early 2016, we were able to sync up with several of the main FPGA vendors at the GPU Technology Conference (the irony) and co-located OpenPower Summit, where we heard quite a bit about the golden age of the FPGA—all brought about by the cloud. With an estimated 75 percent of all servers being sold to live a virtualized life, the market rationale is not difficult to see—but performance per watt is the real story, especially when compared to GPUs, says Mike Strickland, who directs the compute and storage group at Intel/Altera. That puts Strickland in direct contact with HPC and hyperscale shops and gives him an understanding of their architectural considerations.

Although FPGAs have the reputation of being expensive, at high volume they are on par with other accelerators, Strickland explained, pointing to Microsoft as a key example. However, he says that the efficiencies of the performance boost far outstrip GPUs for neural algorithms, which leads to additional savings. There are numerous charts and arts highlighting the price/performance potential of FPGAs in both bare metal and virtual environments, but the real question is that stubborn fourth "P' – programming.

There are programming parallels that make the possibility of an FPGA boom more practical. Strickland estimates there are around 20,000 CUDA programmers in the world, which he says demonstrates the size of the potential OpenCL-based approach to coding for FPGAs. The CUDA and OpenCL models are quite a bit more similar than they have been in the past, but both accelerator programming frameworks come with a reasonably large learning curve. For developers to branch out to either GPUs or FPGAs means they must see the potential for big performance, efficiency, and other gains—and that's the message the FPGA world is trying to push with its focus on deep learning and neural networks.

It is not unreasonable to see how key advancements might lead to FPGAs as a service in, for instance, the Amazon cloud. There are already GPU instance types available, which one could argue might lead to more testing and development with CUDA code for new workloads. For Altera or Xilinx to find their FPGAs offered on IaaS clouds could encourage more OpenCL and programming progress, and might prove to be an ultra-efficient accelerator addition for cloud providers hoping to provide users with a boost without high power and heat complications. We will talk more about this in a coming chapter that discusses how Amazon has sought to integrate FPGAs into its own future cloud offerings—and what it means to some companies whose software depends on FPGA acceleration in large-scale analytics and gene sequence analysis.

Without a simpler way to run complex deep learning and neural network code, all the potential power and acceleration boosts are lost on the market. During a presentation at GTC and the OpenPower Summit in early 2016, Manoj Roge, who directs the datacenter division of Xilinx, said that FPGAs stand to make gains in the near future for specialized workloads. This has always been the case (there are a few places where FPGAs do really well, on Monte Carlo simulations for instance), but the cloud is making access more practical and helping users onboard faster.

"We are in the age of software defined everything—to virtualize all elements of the datacenter from compute, storage and networking and deliver it as a service or cloud," Roge said. "A lot has gone into virtualizing compute and storage, not as much on the networking side, but that's where standards and a robust ecosystem come into play. There's a need to build things with standards but some, including standardizing on X86, are not good for all workloads. Some are seeing how they can get speedups for specific workloads with FPGAs and GPUs."

The challenges, even for software-defined datacenters, still boil down to power and thermal density, something that multicore processors sought to tackle. "We need to rethink datacenter architecture so we can boost performance and reduce latency. The answer is heterogeneous architecture for specialized workloads." The clearest path to accessing that future? To some, it appears to be via the cloud.

CHAPTER TWO

An FPGA Future in the Cloud

In the conference call announcing the deal for Altera, Intel CEO Brian Krzanich said that up to a third of cloud service providers. Intel's plan is to get a Xeon processor and an Altera FPGA on the same chip package by the end of 2016 and ramp its production through 2017, with a follow-on product that actually puts the CPU and the FPGA circuits on the same die in monolithic fashion shortly after that.

The workloads driving that kind of optimistic adoption are set to be varied, but with advances in programmability and capability, these estimates might be right on target.

While Intel might have stolen headlines at the beginning of 2016, the year ended with rival FPGA maker, Xilinx making big waves on the cloud front with its devices being outfitted in the Amazon cloud as part of a much larger planned program to open access in 2017 and beyond.

The solution to the adoption problem, aside from some key programmability steps we will describe later, has been to make FPGAs available in a cloud environment—something that also helps reroute around the cost for those looking to experiment. As obvious of a move as this seems to be, the large cloud providers have been slow to make this happen. That is changing too, but we still have some time to go before both the hardware and tooling are on the same cloud platform for a larger range of developers and users.

FPGA makers are certainly seeing the writing on the wall when it comes to their devices being paired with big public cloud instances. When Intel acquired Altera last year, the question was what it might mean for the swiftly expanding market for reconfigurable computing and more narrowly, what it could signal for the other leading FPGA company, Xilinx. It was clear well in advance of the Intel acquisition that FPGAs were poised to make greater inroads in the datacenter, a matter that was confirmed by Intel's figure of massive FPGA cloud use, which were figures they used to prop up the ultra-high $16.7 billion acquisition sum (for a company that saw itself playing in a future market worth around $1 billion).

While Intel may have been banking on hyperscale and cloud companies to support their FPGA investments, standalone rival, Xilinx bolstered its efforts to reach out to larger markets for both the application and networking/ storage sides of its FPGA business. In that meantime as well, a great many new efforts have cropped up showing how FPGAs can snap into an ever-widening array of workloads on the compute side, particularly as machine learning, IoT, and other trends continue to ramp.

In short, it has been a good time to be the underdog, if we can call Xilinx that for being on its own. Just before the close of 2016, Amazon Web Services announced the very thing Intel had been banking on—a forthcoming host of FPGA-enabled nodes on its EC2 cloud. This offering will be leveraging Xilinx devices, which as the FPGA maker's SVP of Corporate Strategy, Steve Glaser tells *The Next Platform*, shows FPGAs going mainstream in hyperscale datacenters. "We recently introduced the Xilinx Reconfigurable Acceleration Stack to speed up this type of adoption and the AWS announcement is further evidence this is happening right now and the momentum is building."

"The Amazon EC2 FPGA instances (F1) program for now includes the FPGA Developer AMI and Hardware

Developer Kit with everything a developer needs to develop, simulate, debug, and compile hardware acceleration code. Once the FPGA design is complete, developers can save it as an Amazon FPGA Image (AFI) and deploy it to an F1 instance and bring their own FPGA designs, or go to the AWS Marketplace to find pre-built AFIs that include common hardware accelerations. FPGAs are connected to F1 instances through a dedicated, isolated network fabric and are not shared across instances, users, or accounts."

As the company's Jeff Barr describes, these nodes can have up to eight of the following in a single F1 instance: the 16nm Xilinx UltraScale+ VU9P, dedicated PCIe interface with the 2.3 GHz base-speed Broadwell E5-2686 CPU, and four DDR4 channels.

Barr says that "in instances with more than one FPGA, dedicated PCIe fabric allows the FPGA to share the same memory address space and to communicate with each other across a PCIe fabric at up to 12 Gbps in each direction. The FPGAs within an instance share access to the 400 Gbps bidirectional ring for low-latency, high-bandwidth communication" although as one can imagine, this will take protocol writing to make happen on the user end.

Users can write their code using either VHDL or Verilog and then use verification tools from Xilinx, including their Vivado design tools or other compilers. OpenCL tools were not described, we will follow up on that when we can get comment from AWS. So far, the tooling is the lower level stuff for experienced users and as a side note, is only open in the AWS US East region for now. On that code note, we have to temper the idea that this availability will propel FPGAs into the real mainstream because so far, the tooling AWS is providing is only really going to appeal to folks who already have experience working with FPGAs, even if Jeff Barr says it is simplified. It is for this reason that those calling this

"FPGA as a Service" might be a bit off track since that would imply the tooling is in place to make it a real out of the virtual box option.

It is difficult to tell just how many nodes inside AWS datacenters will be FPGA enabled for user adoption, but the beauty about these devices is that AWS might have already had these installed on EC2 servers anyway to support its so-called "smart NICs" which use reprogrammable logic for network activity, but leave a good part of the FPGA idle for other purposes in the meantime and can come with algorithms pre-loaded that can be chewed on for whatever part of the workload desired.

This part is just speculation since AWS has not responded to our requests for more information, but this is something Microsoft does with its FPGA "personalities" that can handle SDN activity then flop over to handle Bing and other workloads they proved out with their Catapult servers. These Catapult machines, by the way, have given way to the Olympus servers (a rack version with full-bore PCIe cards that can handle GPUs, FPGAs, or a mix) that we can see Microsoft spinning out to offer in Azure by the same time AWS F1 (FPGA) instances are made widely available (the announcement today was a developer preview of the offering).

"Large-scale financial clearing risk management is essential to delivering value for our customers," said Kevin Kometer, Chief Information officer, Chief Information Officer, CME Group. "CME Group has long been an innovator in the use of accelerated computing, for the clearning risk management of increasingly complex instruments, including extensive research into FPGAs. Amazon EC2 F1 instances will allow us to substantially accelerate rate of innovation of risk analysis for our customers, while delivering greater cost efficiency relative to using traditional IT infrastructure."

While Amazon may be the first major cloud to offer FPGAs in the cloud (with Azure right on its heels, we imagine) they are not the first overall. Some of the smaller high performance computing cloud companies have been early to FPGA bat, including supercomputing cloud company, Nimbix.

What Nimbix has done that AWS hasn't, at least at this point, is put the higher-level developer tools and environments in place that will actually open FPGA adoption to a much larger potential market. It could be they are testing the waters to see how interested users are first before investing in the richer set of interfaces and tools to talk to FPGAs while leveraging the same parts they use for other datacenter functions, but if interest in FPGAs even just on the application front side is any indicator, there is a rich opportunity here. We will talk more about Nimbix and FPGAs for the supercomputing set shortly, but the point remains the same around tooling.

While the focus has been on how this has been an important boost for Xilinx, just because AWS didn't pick Altera for this particular doesn't mean the Intel-led company has been left on the cutting room floor. While Xilinx will likely continue picking up deals like this one (which is good, otherwise Altera/Intel will have a monopoly situation), it puts the two FPGA makers to the test—and keeps them on their development toes. Although here we tend to cover the application and to a lesser extent, network/storage functions and futures of FPGAs, there is a wide world of FPGA device applications in embedded, military, IoT, and other markets that is also on the grow.

Amazon Web Services might be offering FPGAs in an EC2 cloud environment, but this is still a far cry from the FPGA-as-a-service vision many hold for the future. Nonetheless, it is a remarkable offering in terms of the bleeding-edge Xilinx accelerator. The real success of these FPGA (F1) instances now depends on pulling in the right partnerships and tools to snap a larger user base

together—one that would ideally include non-FPGA experts.

In its F1 instance announcement, AWS made it clear that for the developer preview, there are only VHDL and Verilog programmer tools, which are very low-level, expert interfaces. There was no reference yet to OpenCL or other hooks, but of course, given such a high-end FPGA, AWS might have strategized to go for the top end of that user market first to give access to a Xilinx part that has yet to hit many datacenters. Also, this is an early stage effort designed to appeal to existing FPGA users who might not have otherwise had access to the brand new 16nm UltraScale Plus FPGA.

What we found out after talking to companies that carve a niche by offering such interfaces via compilers, tools, and frameworks (often on their own appliances) is that they are a key to AWS's strategy to onboard new users. On the analytics side, Ryft, an FPGA accelerated appliance maker for large-scale analytics is a good example, whereas for domain-specific FPGA companies, including Edico will help usher in new F1 users for genomics research. In short, the strategy for AWS is a familiar one; partner for the expertise, which lends to the customer base, which creates a stronger cloud product (and presumably, more options for the partner due to increased reach).

Ryft's FPGA accelerated analytics business is driven by government, healthcare, and financial services with an increasing push on the genomics side for the types of large-scale pattern matching required in searching for gene sequencing anomalies. The company says their FPGA-boosted analytics on the AWS F1 instances to bolster elastic search will allow for such searches across multiple datasets, both for individuals and entire populations. "It's very difficult with conventional analytics tools using just CPUs or even GPUs to this type of search effectively. FPGAs are a natural fit for that, and since this is a market the cloud providers want to tackle, it's in their interest to keep pushing the envelope," McGarry says. The company's analytics engine bypasses the standard ETL processes for this and other workloads, and this will keep speeding workloads this and the others powering their business, he adds.

The reason Ryft's business worked at all is because first, the addition of FPGA acceleration workloads like those listed above is not to be dismissed. However, also not to be overlooked is the complexity of using low-level tools to get to the heart of reconfigurability's promise. Like the very few other companies out there providing FPGA acceleration for key workloads, their emphasis is on ultimate abstraction from the hardware and a focus on key workloads. Amazon, it seems, is following their lead, albeit by tapping those who do it best.

As a side note to that domain-specific approach to pulling in non-FPGA experts to AWS F1, we spoke at length with Edico Genome in the wake of the F1 announcement. They use custom FPGA-based hardware to support genomics research, including providing end-to-end sequences in minutes (more detail on that coming in a detailed story next week). McGarry's point about AWS and cloud providers hoping to tackle the genomics boom with enriched platforms for sequencing and analysis is an important one, and could help explain why AWS is getting in front of the FPGA trend. In the announcement of the F1 instances, genomic research was at the top of the potential use cases list.

Ryft is jumping in with AWS to boost elastic search (especially useful for the genomics use case) following a nine-month collaboration with the cloud giant to share insight about how to make FPGA-based analytics workloads hum. "Amazon certainly understands that the success of this instance is dependent on their ability to allow people to abstract the complexity of FPGAs; to provide the interfaces and an existing analytics ecosystem for this and other instances," Ryft CEO, Des Wilson, tells *The Next Platform*. "They came to us because of our ability to do that and they will move this into the mainstream this way."

AWS has its complexity level set high with the Xilinx Ultrascale 8-FPGA nodes it has designed for the F1 instance, but according to Ryft's engineering lead, Pat McGarry, they picked the right part for the times. While it still isn't clear why AWS picked this without bringing along the higher-level OpenCL interfaces and tools (which are easier to program but don't provide the same level of performance) no one we've talked to seems to be anything but excited such a beefy FPGA (backed with an equally beefy Broadwell CPU along with DSP cores.

"I suspect AWS chose this very new and high-end FPGA because they are trying to tackle classes of problems that none of the older generation FPGAs will be able to touch. It has 2.4 million logic elements, a strong Broadwell CPU and 6500 DSP slices. This eats into territory that used to belong solely to GPUs as accelerators and is a much bigger play for Amazon than just providing FPGA technology," McGarry says. "This will eventually allow for all kinds of new machine learning, AI, big data analytics, you name it, all on one platform. They just have to figure out how to these things in a platform way for other verticals like we've managed to do with FPGA based analytics."

One of the key differentiators other than the logic element capabilities and strong host CPU is that there is a much greater opportunity with these parts for partial reconfiguration. Being able to reconfigure an FPGA on the fly for different analytical workloads is a big opportunity McGarry says, and we will continue to see others doing this while also offloading critical network and other datacenter functions on the same device, although Ryft folks were not convinced this is what AWS is doing with their FPGAs. "These are really designed just for the F1 instance, we're quite sure," McGarry stated. "This is purely an EC2 compute play."

Ryft says they were able to provide AWS with key insight into how to solve the problems of distributed accelerators, especially with the highly heterogenous CPU, FPGA, DSP nodes that back the F1 instances. "We had to spend a lot of time working with them to understand their architecture first," Des Wilson says. "We ended up figuring out which pieces of our own architecture for our Ryft One appliances would fit best and use that as a starting point. We had to make the limitations of a cloud provider architecture our strengths and spent months looking at which architectural flaws with that don't match well with FPGAs and use that to our advantage."

Wilson describes the above engineering effort using an example from streaming data to F1 nodes in a cloud environment versus their own hardware. "Instead of streaming data from an SSD, we had to think about doing this directly from S3 or Glacier or some other data source. We took that model and saw as soon as you have that matched with the reconfiguration capabilities of those FPGAs, the AWS architecture lets us connect all that together because of the ring topology with 400 Gb/S of throughput between the FPGAs. It's possible to segment jobs nicely. These are very new FPGAs and while latencies are an issue with many of these tied together, for most of the commercial workloads, this shouldn't be a problem," Wilson says. "This is a big deal for future workloads on FPGAs."

This all begs the question about what this AWS work will mean for Ryft's niche FPGA appliance business, of course. While indeed, they get to pick up potential new users of their analytics packages (an important benefit), they had to share some of their secret sauce with AWS to make such a partnership practical. Ultimately, Wilson says this is a very good thing for their business because there will always be customers who need the ultra-low latency of the Infiniband-connected appliances they sell but the real value is the FPGA analytics software and services. As we know, there are razor-thin margins for anyone in the hardware game, so losing out on this business isn't as devastating at it might otherwise sound.

"Everyone loves their own hardware design, so it was a challenge for us in this F1 work to not redo everything we've done in our architecture and appliance on AWS's own infrastructure. We had to think carefully about their networking architecture, separation of the data across it at scale, and the latencies associated with doing so. Once we got past this, things moved along well."

As we will explore in greater detail next week as we look at how AWS is backing into the FPGA business by using domain-specific companies like Edico Genome, we will look at more of the technical and interface challenges for truly democratizing FPGAs via a cloud model. As we noted in the past, there are companies like Nimbix, that have been working with HPC application developers to integrate FPGA acceleration, but there has not been such a big public push until the AWS announcement.

Since we expect that others, particularly Microsoft, which has deep experience with FPGAs, to announce similar offerings in the near future (an educated guess, of course), looking at how an expert-required device like this will filter to the mainstream through higher-level or domain-specific sources will be interesting–and could spell out how Altera/Intel attack the market when a future integrated part emerges.

Edico Genome is another company that is showing how domain-specific FPGA expertise can be expanded by public cloud availability of reconfigurable devices.

Just one year ago, Edico says its own FPGA approach sequence analysis allowed the time to be pared down to twenty minutes for a single genome using their custom-tailored FPGA-accelerated "Dragen" systems. That was an impressive feat then, but the company's Gavin Stone now says they're pushing near-real time for analysis. "We are able to do this at the speed of the data," he tells *The Next Platform*, saying that the key to this speed is a mix between their own algorithmic tweaks and partial reconfiguration with both Xilinx and Intel/Altera FPGAs.

"Having eight of these FPGAs in a server is a big deal for us. As fast as you can move the data around we can analyze the genome, which is something that took weeks a few years ago, then days not long ago, and now is within minutes to nearly instantly." This speedup now means processing elements, whether the FPGA or CPU (in the case of the F1 instances, there is a beefy Broadwell attached) is freed up for more complex analysis. "The current algorithms have an adequate level of accuracy, but because of computational and practical limits, there were many algorithms we've had in development that could deliver far higher accuracy numbers that previously realized. With FPGAs like this available, we are going to find and develop ways to make current gene analysis even better," Stone adds.

The company's current FPGAs have around a million logic elements, but the newest Xilinx UltraScale parts in the F1 instance have 2.4 million. This is a big boost for genomics workloads, Stone says, and one that will allow teams to maximize the real estate available on the FPGA for far more complex workflows. Of course, to do this means taking a leap in the partial reconfiguration zone—swapping in and out different elements onto the FPGA for specific parts of the workload. This is not a simple task technically but it is the key to getting the full performance and efficiency out of a reconfigurable device.

Partial reconfiguration capabilities have always been available on FPGAs but it has been very difficult to make use of, in part because of the timing challenges with pulling parts in and out with a "live" FPGA that's running other operations. "People have been using this approach but only in small elements; maybe carving out 5% or 10% blocks to swap. We are doing this at a 90% swap level, keeping things like the PCIe controller, drivers, and other essential functions alive and then swapping in other engine blocks on the fly." With the genomics pipeline, for example, there is one block to handle compression, another for mapping and aligning, another for variant calling and so on. Ultimately, that single device can be used fully as an accelerator for many different functions, leading to a far faster time to result—something that is a key point of differentiation for users at genomics centers that want to provide analysis results at the point of care.

"With partial reconfiguration, there is no part of the device that lays dormant. It's technically challenging, but once it works, it works extremely well. The barriers to getting to this point have been huge and we've worked with Xilinx closely on this, but this is becoming more mainstream as people start to realize how to truly use the FPGA to its fullest," Stone says. This work has paid off for Xilinx and end users in other areas, who will find it easier to make use of partial reconfiguration—something that could promise an even larger field for FPGAs to play in.

Although partial reconfiguration is a challenge for many shops, this is the sweet spot for Edico Genome and the reason they can deliver such fast results. That capability will be coming to the cloud soon, albeit hindered slightly by the data movement delay. This is not something that will add a huge barrier, and Stone says this will keep the company's own hardware business alive since there will always be centers that need the ultra-low latency time to results on site.

When Edico's business started, the idea had been around custom ASICs, but the volume and flexibility story wasn't there. Stone says he expects this ASIC versus FPGA question will be less pressing as those in genomics and other areas realize that even though using partial reconfiguration is still not simple, it beats the economics of driving a chip to production and taking such a big financial risk. "We are truly able to use the FPGA to its fullest," says Stone, pointing to the benefits of their partial reconfiguration approach. "We do reconfigure on the fly, in the middle of the workflow, which not many people do. Normally, you have to do a full reconfiguration, but we're keeping a lot of the FPGA live and swapping portions in and out. There is no way we could do that with an ASIC. And as many see now, genomics is evolving quickly; the algorithms change and update and pushing those through quickly with an ASIC would not be feasible." Stone says that the costs of FPGAs will come down as will some of the programmatic complexity, making ASICs make less sense.

"There is widespread adoption of FPGAs now; it's really caught fire over the last few years as we've seen the writing on the wall with Microsoft's Catapult project and others getting a lot of attention. There used to be niche providers in the cloud but with Amazon putting FPGAs out there, it is going to more mainstream—not just for genomics but other data-rich applications."

Other companies are taking advantage of FPGAs in the cloud, although with a more tailored approach. Even though Amazon might be the first major public cloud player to adopt FPGAs for larger use, others have been at this game longer. Take, for example, Nimbix—a supercomputing-focused cloud that delivers high-end hardware to users in high performance computing (HPC) and more recently, deep learning and machine learning. Nimbix saw the opportunity for cloud-based FPGAs and tooling several years ago and can shed some light on where users at the high end are going with FPGAs delivered remotely.

Back in 2010, when the term "cloud computing" was still laden with peril and mystery for many users in enterprise and high performance computing, HPC cloud startup, Nimbix, stepped out to tackle that perceived risk for some of the most challenging, latency-sensitive applications.

At the time, there were only a handful of small companies catering to the needs of high performance computing applications and those that existed were developing clever middleware to hook into AWS infrastructure. There were a few companies offering true "HPC as a service" (distinct datacenters designed to fit such workloads that could be accessed via a web interface or APIs) but many of those have gone relatively quiet over the last couple of years.

When Nimbix got its start, the possibilities of running HPC workloads in the cloud was the subject of great debate in the academic-dominated scientific computing realm. As mentioned above, concerns about latency in the performance-conscious realm of these applications loomed large, as did the more general concerns about the cost of moving data, the remote hardware capability for running demanding jobs, and the availability of notoriously expensive licenses from HPC ISVs.

While Amazon and its competitors plugged away at the licensing problem, they were still missing the hardware and middleware specialization needed to make HPC in the cloud truly possible, even those AWS tried early on to address this by adding 10 GB Ethernet and multicore CPU options (and later, lower-end Nvidia GRID GPUs). In those early days, this difficulty is what fueled the rise of other HPC cloud startups like Cycle Computing, which made running complex jobs on AWS more seamless—but the other way to tackle the problem was simply to build both the hardware and software and wrap it neatly in a cloud operating system that could orchestrate HPC workflows with those needs in mind.

This is the approach Nimbix took and they quickly set about adding unique hardware in addition to building their JARVICE cloud operating system and orchestration layer, which is not entirely unlike OpenStack. The custom-built JARVICE platform sits on top of Linux to allow it run on the heterogeneous collection of hardware that sits in a distributed set of datacenters in the Dallas metro area (with more planned soon, including in Europe and Asia). This manages the clusters and workflows, assigns resources, and manages the containers that power user applications.

Leo Reiter, CTO at Nimbix tells *The Next Platform* their typical users fall into two categories. On the one hand there is the bread and butter simulation customer that users the many solvers and applications in the Nimbix library of scientific and technical computing applications they have license agreements with. For these users, they provide the data and performance parameters and the system orchestrates the workflows using JARVICE and their container approach to application delivery. Counted in this group are other users with high performance data analysis or machine learning needs. On the other end are their developer users, who can use Nimbix as a Paas to deliver their own workflows or applications and stick those in the public or private catalog. Of course, to do all of this with high performance and scalability means the Nimbix folks had to give some serious thought to hardware infrastructure.

Nimbix has been providing Xilinx FPGAs in their cloud since 2010 for researchers and the Xilinx development team, but they also have a wide range of Nvidia GPUs—from the low end Maxwell-based parts for the Titan X (for machine learning training) to the new M40 processors for deep learning all the way up to the Nvidia Tesla K80 cards for those with high performance simulations and analytics. Much of the processor environment consists of 16-core Haswell parts, which they can create secure, fractional nodes from as needed (making a 16-core part look like a 4-core with the necessary memory apportionment, etc.). They also use Infiniband for all nodes and for their storage system. So far, their cloud compares only to some elements Microsoft has integrated (they now have some K80s and Inifiniband capabilities) but overall, Reiter says, they are succeeding because no other cloud provider is making the hardware investments to quite the same degree. He points to the fact that there are GPUs on AWS, but the Grid parts aren't meaty enough to handle the seismic, bioinformatics, engineering and other HPC oriented workflows—and even for deep learning training these are insufficient to their users.

What is interesting here is that just as companies that have specialized in HPC hardware are finding their gear is a good fit for deep learning training and broader machine learning applications, so too is Nimbix finding a potential new path. They have managed to carve out a niche in supercomputing and a few other areas, but so far, there aren't a lot of robust, tuned high performance hardware options as a service that fit the machine learning bill. We noted that Nervana Systems (recently acquired by Intel) is doing this, and there are a few others who are offering deep learning as a service, but a company that HPC users might know might be very well positioned as deep learning and HPC merge in some application areas and require a remote sandbox—or eventual production environment.

Reiter says they are seeing more interest in deep learning and machine learning and have added robustness to their software stack with hooks for TensorFlow, Torch, and other frameworks. Since they already have the heterogeneous hardware on site and a proven business model behind them, we could see Nimbix move from quiet company from the research regions to HPC push into greater visibility via a new crop of machine learning applications and end users.

These two areas; high performance computing and deep learning/machine are where we see big opportunities for FPGAs in the next few years. These emerging workload demands, along with different delivery models (cloud, on-prem) bolster the belief that the future is bright for FPGAs on the application acceleration front.

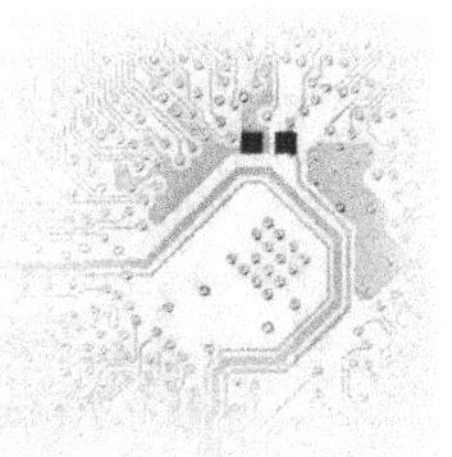

CHAPTER THREE

FPGAs in High Performance Computing

So far, we have established a potential path to FPGA adoption in cloud, hyperscale, and some enterprise applications (with machine learning next on our list). But what about supercomputing, or high performance computing, which seems like a fit for the power-aware capabilities of an FPGA?

At the last five annual Supercomputing Conferences, an underlying theme has been the potential of accelerators to add energy-efficient performance. In supercomputing, accelerated computing is becoming the new normal—and there is a distinct possibility that FPGAs could find a place in the coveted Top 500 list of supercomputers in the future, especially when integrated CPU and FPGA parts begin to appear.

The trend line for large-scale datacenters on the hyperscale end and for some specialized commercial and research HPC-centric workloads in genomics, seismic analysis, and finance, intersects with where FPGAs are heading. A lack of FPGA accelerated systems on the Top 500 now is no indication of what the future holds. With Intel aiming to eventually integrate the Altera FPGA IP they acquired this year for over $16 billion and the other major FPGA maker, Xilinx, beefing up its hardware and software approach to broader markets, we are poised for a shakeup in accelerated computing at extreme scale.

Intel is not the only company that has a path to integrated FPGAs with CPUs and the companion software environment to wrap around it. Armed with a new Qualcomm partnership for low-power FPGA approaches to suit hyperscale datacenter needs, and a freshly announced partnership (that is finally formalized after several years of co-development) with IBM's Open Power Foundation, Xilinx sees a path to the large-scale datacenter. High performance computing applications are part of this roadmap—but so too are workloads that other accelerator approaches are tackling, including Nvidia with its newly launched Tesla M40 and M4 processors aimed at hyperscale datacenters running machine learning training and inference workloads, among others in data analytics, security, and beyond.

There is reason to believe that within the next few years, there could be at least a few entrants on the Top 500 list that are taking advantage of FPGA acceleration as well. With Intel's acquisition of FPGA maker Altera earlier this year and their projections for integration with Xeon CPUs and the other major FPGA company, Xilinx, striking up partnerships with Qualcomm on the low-power ARM processor side and IBM's OpenPower Foundation on the other, the FPGA space is set to become varied and heated enough to spur unprecedented activity for FPGAs on the compute side.

The accelerator story for top supercomputers is a strong one, starting with GPUs, which were snapped in as coprocessors on some of the world's largest systems over the last five years. Since then, other accelerator options, including the Xeon Phi (and next generation Knights Landing coming soon) have emerged, which aim to provide a more programmable interface to accelerate HPC applications.

Even though there are no systems sporting FPGAs on the Top 500 list, there is some experimental work happening in traditional HPC applications using programmable gates, coupled with on-chip and programmatic enhancements, which could make FPGAs more attractive for HPC. One of the major barriers for supercomputing or general enterprise datacenters is simply that these devices are difficult to program, however. We described how that is changing with some recent momentum around moving FPGAs closer to procedural languages and OpenCL, but it is still a long road.

The good news for HPC, especially in research and academic environments, is that there are plenty of graduate students on-hand at the national labs and universities that can spend the time needed to specialize in FPGA development and programming. But still, that is not enough to make these devices ready for the next generation of supercomputers, even with momentum from vendors like IBM, who through their CAPI interface and shared memory, are making FPGAs and other devices closer to the compute and eventually, more integrated programmatically.

Unlike some enterprise workloads, HPC is historically floating point intensive, which means that these applications are not a good fit for the FPGA, at least until vendors can snap in dedicated floating point units—something that Altera has publicly said they are working on for future generations (and Xilinx will likely follow suit). It is not that it's impossible to get reasonable floating point performance off FPGAs now, but there are a lot of gates and it would be woefully inefficient to do so. GPUs, the dominant accelerator in HPC, are stuffed with floating point units, on the other hand and coupled with their rich CUDA (and OpenCL) ecosystems, are still the simpler choice for acceleration.

FPGAs do offer the fine grained parallelism and low power consumption of other accelerators, with extreme configurability added in. But beyond the floating point limitations to date, the difficult programming environment, there is also another big limitation—albeit one that will be overcome soon enough. For these applications, FPGAs are limited by the internal memory on the chip. The bandwidth might be great, but without enough memory, this is a big limitation.

Wim Vanderbauwhede at the University of Glasgow's School of Computing Science has been working with FPGAs for well over a decade and has moved his research into the area of looking at key HPC applications and how they match to FPGAs. In a chat with *The Next Platform* in 2016, he talked about how for things like search and working on large graphs, the FPGA is well-suited (although in the latter case, memory limitations are still an issue).

There are already a range of high performance computing applications that can be run on the FPGA for a sizable boost, presuming the code legwork can be done. According to Vanderbauwhede (whom incidentally, wrote the book on high performance computing on FPGAs) if code has already been optimized for other accelerators, including GPUs, much of the heavy lifting has already been done. In his teams' work on FPGAs for a select set of HPC applications, it took about one month for a full time person to prepare code to run at high performance and efficiency on FPGAs—a boost that is worth the effort in areas like financial services where stock option pricing and Black Scholes models really let the FPGA shine. In this example, along with others that are multi-kernel and deal with relatively small datasets (including molecular dynamics and key biomedical applications) FPGAs can perform well, but there are other areas that offer opportunity in the future that are worth picking through for now, including weather modeling—an interesting target since it is straight number crunching and memory hungry.

More recently, Vanderbauwhede and his team have taken a traditional HPC application in weather modeling to look for speedups with FPGA acceleration. While they have had success performance-wise, he says that for these centers and other large supercomputing sites, it's far more a matter of performance per watt. This is where FPGAs will find a fit in HPC while the memory and programming environments catch up.

"There is use for some HPC applications like weather modeling as long as you are able to use the full parallelism of the device. What we have done shows it is just as power efficient as a multi-core system and far more efficient than any original single-core code they used to run. But of course, memory is still an issue."

Vanderbauwhede says that the problem with FPGAs for weather simulations and a select set of other HPC application is that there is always a limitation in the number of gates one has to work with. That means, if you can split it up and reconfigure the FPGA to do different parts of the program at different times, and it is fast enough, the cost of swapping the configuration is offset—assuming, of course, there is enough data to work with. "So in the weather example, the time spent computing a volume of atmosphere will be larger than the time it takes to reconfigure the device. That's exciting because

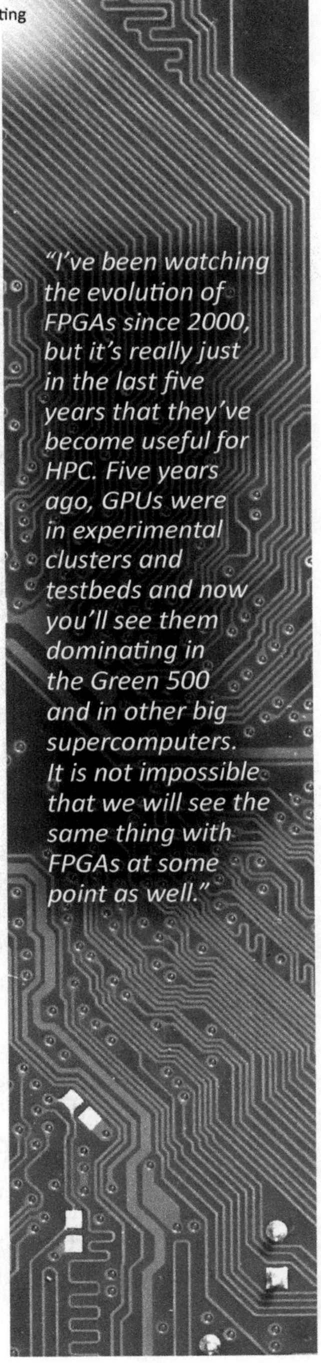

"I've been watching the evolution of FPGAs since 2000, but it's really just in the last five years that they've become useful for HPC. Five years ago, GPUs were in experimental clusters and testbeds and now you'll see them dominating in the Green 500 and in other big supercomputers. It is not impossible that we will see the same thing with FPGAs at some point as well."

until recently, even though they're called reconfigurable devices, a lot of that didn't happen on the fly."

In terms of future directions for FPGAs in HPC, Vanderbauwhede agrees that once they are more tightly integrated and can share memory it would be a step change. "If you have an FPGA in the socket where normally a processor would sit so that it has access to the front side bus, that is a game changer. At the moment, the big problem is that it's a PCIe offload model. There's so much data that needs shifted back and forth, so getting around that will likely open FPGAs for more users.

FPGAs might be the next big thing for a growing host of workloads in high performance and enterprise computing, but for smaller companies, not to mention research institutions, the process of onboarding is not simple—or inexpensive.

From the systems themselves to the programming and compiler tools, even experimenting with FPGAs is an undertaking resource-wise. While there are efforts to lessen this load, including the addition of a richer OpenCL toolset and cloud-based FPGA offerings, the ecosystem for FPGAs is still developing, which means getting to "point A" is where some foundational work needs to be done. We have already seen a great deal of momentum for machine learning codes that can take advantage of FPGA-based systems, but despite the increased interest in FPGAs at the annual Supercomputing Conference in Austin, there is still relatively little information about which scientific codes might be a good fit for FPGA acceleration.

The dominant accelerator in high performance computing is the GPU, with a swell of scientific codes across a wide range of disciplines available and CUDA-ready. This type of ecosystem development for HPC has not happened to any great degree among the FPGA vendors, in part because the compute market has been a secondary focus. There are other technical requirements (we'll get to that in a moment) that have not made HPC codes well-suited to FPGAs, but a lot of that is set to change. The goal now is to offer researchers a rare chance to actually experiment with their codes using a real system—something that is hard to do "in the wild" without some serious up-front cost and expertise.

For Derek Chiou, former professor at the University of Texas at Austin and now full-timer at Microsoft on the Catapult project, this lack of research access to FPGAs is a problem—but there are efforts to bring FPGAs to the academic masses.

For instance, the Texas Advanced Computing Center (TACC), has set its sights on the future of FPGAs by making the Altera Stratix V FPGA-outfitted Catapult servers available for free for researchers, assuming of course, they are willing to share the code they develop on the platform. Chiou, who started his work with FPGAs as a post doc at MIT, toiled away on router architectures at Avici Systems before becoming a professor at UT Austin, now is a liaison between TACC and Microsoft. The effort now counts several hardware partners, including both major FPGA makers, Xilinix and Altera, and has commitments for tooling from Intel, compilers from Bluespec, and a C-to-gates compiler from Impulse Accelerated Technologies. The TACC work is supported by an National Science Foundation grant and builds on work Chiou and teams did on other projects, including a program called Research Accelerator for Multiple Processors (RAMP) that examined the problems of building ultra-dense, multicore systems from FPGAs and other processors.

TACC will have an impressive array once the systems are up and running. Currently, the project is in the install phase, but Chiou says there will be 400, or perhaps slightly more, machines. They are in the process of evaluating codes for the systems now, but since this is the largest open installation of FPGA-based systems, he expects there to be great demand.

The real takeaway from the work at TACC with FPGAs will be seeing the ultimate scalability of the system across a new range of codes. The configuration chosen for TACC is a miniature version of what Microsoft implemented for its Bing search engine, which was able to significantly accelerate Bing across a cluster of 1,632 servers featuring the same Stratix V in a half-rack of 48 nodes with one single FPGA board connected via PCIe and connected to the other FPGAs via SAS cables.

It will be interesting to see how many scientific codes find a fit with the FPGA systems since the Altera Stratix V does not offer native support for floating point. That is not a deal-killer since soft logic can be used as a workaround, but that is not very efficient. Once Altera rolls out its Arria 10 FPGAs, however, a new world could open for high performance computing codes since those will include hardened 32-bit floating point units. The Arria 10 has on the order of 1.5 teraflops of floating point performance, which is quite good considering the low amount of power they burn. The next generation Stratix 10 is expected to have 10 teraflops of 32-bit floating point performance. While 64-bit is the preference for many scientific and technical codes, but there Chiou says there are several codes that will hold nice nicely to 32-bit and given the performance per watt, which is better than GPUs and definitely CPUs, that tradeoff might be a consideration.

Following Chiou's work on the Catapult project to accelerate Bing, the team noted that the big challenge ahead still lies in programmability. While he says that RTL and Verilog are common tools for his work, efforts being made in Scala and OpenCL represent a path forward for wider adoption. "Longer term, more integrated development tools will be necessary to increase the programmability of these fabrics beyond teams of specialists working with large-scale service developers." Within the next decade to fifteen years, bringing us to the end of Moore's Law, however, "compilation to a combination of hardware and software will be commonplace."

The most compelling applications and system research will be given an allocation and consulting services to port their applications to Catapult to make effective use of FPGA-based acceleration, all at no cost.

Oftentimes, the work done at the ultra-high end (supercomputing) trickles down to the enterprise, but for FPGAs, the situation is somewhat flipped. The hyperscale web companies and social networks are the ones leading the charge toward FPGA development in large part and much of this begins with deep learning and machine learning.

Research has been ongoing to integrate FPGAs into existing HPC workflows in several key areas. Among several examples we wanted to highlight are those from high energy physics hub, CERN, as well as among the supercomputing set at weather modeling centers.

At the core of the rise in interest in FPGAs in HPC, at least for the non-specialist in programmable hardware, is OpenCL—a framework that extends the programmatic ease of other accelerators, including GPUs. While it is generally agreed that there is a performance hit created with this added level of abstraction from programming the device directly, it does mean that scientist and other end users might be more willing to give FPGAs a second look beyond the areas where they are already used—often as an option to an expensive custom ASIC.

This story is playing out at high energy physics research center, CERN, which has historically kept an open mind about adopting and integrating diverse accelerator, memory, storage, and other technologies—as long as experiments can be captured and processed faster and with less power. On the processing and acceleration front, the center uses standard X86 processors with the possibility of more ARM parts in the future, as well as GPUs. Custom ASICs and FPGAs have also been in use

"The idea behind implementing a data acquisition (DAQ) system was to explore the possibility of using OpenCL for more than just acceleration. Many of the design elements needed to realize a DAQ system in OpenCL already exist, mostly as FPGA vendor extensions, some of which are going into subsequent versions of the OpenCL specification. However, a small number of elements are missing, preventing full realization of a complete DAQ system but since these elements have simple, feasible solutions, they could also be implemented if the FPGA tool vendors so desire."

at various CERN sites for a range of monitoring, signal processing, and networking tasks.

For the curious, a 2008 paper sets forth the many ways FPGAs have been used at the center, but with coming technology refreshes and upgrades at CERN research sites, one might expect a potentially new set of use cases for FPGAs—specifically as processors for major portions of CERN workloads. With the availability of OpenCL as a gateway, the possibilities for non FPGA experts have opened wide, making FPGAs a source of interest not just for traditional uses at CERN, or even for workload acceleration, but as key processing elements for vital segments of major experiments, including the Large Hadron Collider LHCb experiment.

Following upgrades, this particular experiment will cull 500 data sources, each of which will create data at 100 Gbps. This work presents challenges on both the data acquisition and algorithm acceleration front, which put even state of the art FPGAs to the test, according to Srikanth Sridharan, Marie Curie Fellow at CERN. His team is looking at how to use OpenCL for FPGAs in a way that goes beyond acceleration—all the while leveraging OpenCL to demonstrate its adoptability for the larger sets of domain scientists who have little time to dig into the complexities of hardware description languages and techniques.

Sridharan has spent much of his career working with FPGAs working in industry for companies like Qualcomm as well as large research hubs like the NSF Center for High Performance Reconfigurable Computing (CHREC). Now at CERN, he is turning his attention to the role FPGAs might play in high energy physics, although not in ways one might expect. While his team is focused on acceleration of various algorithms (the area where a great deal of research into GPUs and other accelerators tends to fit) a new idea—using FPGAs not as accelerators, but as low power, high performance data acquisition system processors, is where is most recent work lies.

FPGAs are not a new addition to the data acquisition (DAQ) platform at CERN, but the OpenCL use to program FPGAs for more than acceleration is a new element. Sridharan says they were initially used for this part of the experiment workload to "collate the streaming data coming off the front end electronics over multiple channels." He says they can also be used in the "low level trigger system where the acquired data needs to be quickly processed to arrive at trigger decisions." However, he says that the custom nature of this and the need to do operate in high radiation environments "make any other technology unsuitable for these purpose and ASICs are suitable only for high volume production and are unviable for these applications due to prohibitive costs."

Although the DAQ tests revealed FPGAs using OpenCL as a viable tool, with the few missing pieces he notes above, none of which will prevent future application, algorithm acceleration on FPGA using the Altera compiler for OpenCL revealed scattered results. For one test, FPGAs performed far better and with much better efficiency than the GPUs that were used for this part of the task, and for the other experiment, they were significantly worse. This could be in part to a lack of thorough optimization, Sridharan says, but the team will continue to explore.

Overall, he says an optimized implementation for FPGA on the algorithm acceleration side would provide a better sense, but ultimately, "it cannot be denied that OpenCL makes exploiting FPGAs for acceleration as easy as exploiting GPUs. That is a long way from the days of painstaking efforts to create a cycle accurate HDL design, functionally verifying it, debugging the design errors, and fixing the timing violations to realize a working system." Further, he notes that in cases where optimization work for FPGAs have been done, the performance per watt story is an attractive one for CERN. "Extracting more parallelism from the algorithm, creating an FPGA optimized implementation, investigating the huge drop in performance for some kernels, and also accurate power profiling of the design could be the direction of future work."

"it remains to be seen how such a system would perform compared to a custom implementation in VHDL/ Verilog, but there definitely exists a case for OpenCL in this application due to the massive productivity gain and ease of use it offers," he says. Further, he says that the wider accessibility of OpenCL means non-FPGA experts can design, debug and maintain the code.

Weather modeling and forecasting centers are among some of the top users of supercomputing systems and are at the top of the list when it comes to areas that could benefit from exascale-class compute power.

However, for modeling centers, even those with the most powerful machines, there is a great deal of leg work on the code front in particular to scale to that potential. Still, many, including most recently the UK Met Office, have planted a stake in the ground for exascale—and they are looking beyond traditional architectures to meet the power and scalability demands they'll be facing in the 2020-plus timeframe.

The Met Office released a report in mid-2016 on its own requirements for exascale computing, which follows news from ISC16 last week about the vision for the 2017 systems to back its efforts which focuses on a number of elements for R&D emphasis. These include the exploration of new architectures, bolstering programming models, enhancing I/O and workflow and coupling complex multi-scale and multi-timescale models. The domain specific nature of the code work requires a more focused deep dive, but one of the key elements in the report is the focus on emerging architectures.

As the authors note, "Total power requirements suggest that CPUs will not be suitable commodity processors for supercomputers in the future." In addition to looking to more common accelerators, including GPUs and Intel Xeon Phi, the Met Office is keen on watching 64-bit ARM developments as well as, most surprising, FPGAs as suitable offload engines.

"FPGAs are a well-established technology but difficult for applications developed using high level languages. Two potential research avenues include first developing a software stack to transform high level language code and transform it into hardware logic for FPGAs."

There has already been work to develop the software backbone for FPGAs, but this is the first time we've seen public statements from a major weather center directing attention to it. Among the other novel architectures is the D-Wave quantum computer, which can "in principle, compute the entire space of a minimization problem, albeit with some non-trivial restrictions on how that minimization problem is defined." The Met Office also notes that quantum optical devices are another potential avenue and says "research into how such devices might be exploited to perform simple computations and they can be coupled into existing software stacks and what new algorithms might be possible."

Although the weather modeling and prediction arena might be interested in what comes after Moore's Law, nearly all centers are operating with CPU-only machines. ECMWF and others have some GPUs on their systems, which are used for research versus production (as we understand based on our last check-in with the center) because of the nature of their codes. With some exceptions in the research arena, these codes are complex and not amenable to GPU acceleration except for certain parts that can be offloaded. This is not to say that GPUs will never find a place in weather prediction, or FPGAs for that matter, but if the weather communities are serious about exploring new architectures, investing in the code to fit the next generation of systems will be a requirement.

As is the case in many other scientific computing domains, writing codes from scratch to fit new architectures is not an option. Many codes, including WRF and others, have been developed over the course of many decades, with tweaks to suit the addition of new cores and memory options for optimizations. If integrating GPUs is difficult, one can only imagine the road ahead for FPGAs, which already have the reputation for being difficult to program.

For weather, we can predict more of the same ahead for the pre-exascale machines—and perhaps even those in the 2020 timeframe. In 2016, just four years away from the time when some of the first exaflop capable machines might be announced, there are many Top 500 supercomputers devoted to weather; many of which are in the top 100. Since these centers buy duplicate systems for continuity and research reasons, there are dual machines at ECMWF at the #17 and #18 spots and another duo at the #29 and #30 spot from the UK Met Office, which ran the Top 500 benchmark on its newest Cray XC40 systems for the first time to achieve the high ranking. Other machines in the top 50 of the list include the Korean Meteorological

Administration and their twin systems (also Cray XC40) and NOAA's new Cray systems at the #51 and #52 spots.

The full report from the UK Met Office sheds some light on the higher level code work that needs to be done, but these are to optimize for larger-scale CPU only machines. The question we will be chasing in the coming months is how much effort, both in terms of codes and systems, will be required to get weather modeling out from traditional system architectures in the post 2020 timeframe.

As these and other use cases emerge in HPC, we keep this in mind and turn an eye to a much faster-moving area for development of both systems and software: machine learning and deep learning.

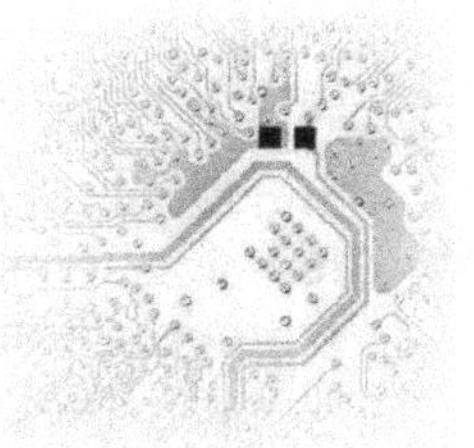

CHAPTER FOUR

FPGA Future for Deep Learning, Machine Learning

It was with reserved skepticism that we listened, not even one year ago, to dramatic predictions about the future growth of the deep learning market—numbers that climbed into the billions some estimates, despite the fact that most applications in the area were powering image tagging or recognition, translation, and other more consumer-oriented services. This was not to say that the potential of deep learning could not be seen springing from these early applications, but rather, the enterprise and scientific possibilities were just on the edge of the horizon.

In the meantime, significant hardware and algorithmic developments have been underway, propping up what appears to be an initial Cambrian explosion of new applications for deep learning frameworks in areas as diverse as energy, medicine, physics, and beyond.

What is most interesting is that in our careful following of peer-reviewed research over the last couple of years, it was only just in Q3 of 2016 that a large number of deep learning applications in diverse domains have cropped up. These breathe new life into the market figures for deep learning that seemed staggering, at best—at worst, woefully optimistic.

These also help explain why companies like Intel are keen to make acquisition for both the hardware and software stacks from companies like Nervana Systems and Movidius, why Nvidia has staked its future on deep learning acceleration, and why a wealth of chip startups with everything from custom ASICs, FPGAs, and other devices have rushed to meet a market that until very recently, just hasn't been present in sufficient volume to warrant such hype. As a counterbalance to that statement, a significant uptick in research employing various deep learning frameworks does not create a market out of thin air either, but the point is that there is momentum in areas of high enterprise and scientific value—and it keeps building.

While cloud represents an overall opportunity for FPGAs and HPC is promising research area, machine learning and the deep learning subset of those problems is another huge opportunity in 2017 and beyond. Major web companies over the course of 2016 put their FPGA machine learning work on center stage, but none made quite as much noise as Microsoft.

After three years of research into how it might accelerate its Bing search engine using field programmable gate arrays (FPGAs), Microsoft came up with a scheme that would let it lash Stratix V devices from Altera to the two-socket server nodes in the minimalist Open Cloud Servers that it has designed expressly for its hyperscale datacenters. These CPU-FPGA hybrids were rolled out into production earlier this year to accelerate Bing page rank functions, and Microsoft started hunting around for other workloads with which to juice with FPGAs.

Deep learning was the next big job that Microsoft is pretty sure can benefit from FPGAs and, importantly, do so within the constraints of its hyperscale infrastructure. Microsoft's systems have unique demands given that Microsoft is building systems, storage, and networks that have to support many different kinds of workloads – all within specific power, thermal, and budget envelopes.

Microsoft's efforts to accelerate the training of various kinds of deep learning approaches – including convolutional neural networks, deep belief neural networks, and recurrent neural networks –put FPGAs through the paces in 2016. Such networks are at the heart of systems that perform computer vision and video and photo recognition, speech recognition and natural language processing; they are also used in recommendation systems that help push products to us in advertisements and on retail sites and behind all kinds of intelligent agent software. These deep learning systems are what gives the modern Internet whatever brains it has, to put it bluntly.

Unlike researchers or commercial enterprises that build deep learning systems that only have to do that one job, Microsoft has to operate its infrastructure at scale, and this makes its choice of accelerator a bit different from operating in the abstract, explained Eric Chung, a researcher in the Microsoft Research Technologies lab that adapted Microsoft's Open Cloud Server so they could be equipped with FPGA accelerators, a system that was called Catapult.

"The datacenter is interesting because we get to scale," explained Chung. "The key here is that we have a large set of fungible resources that we can set up, that we can allocate on

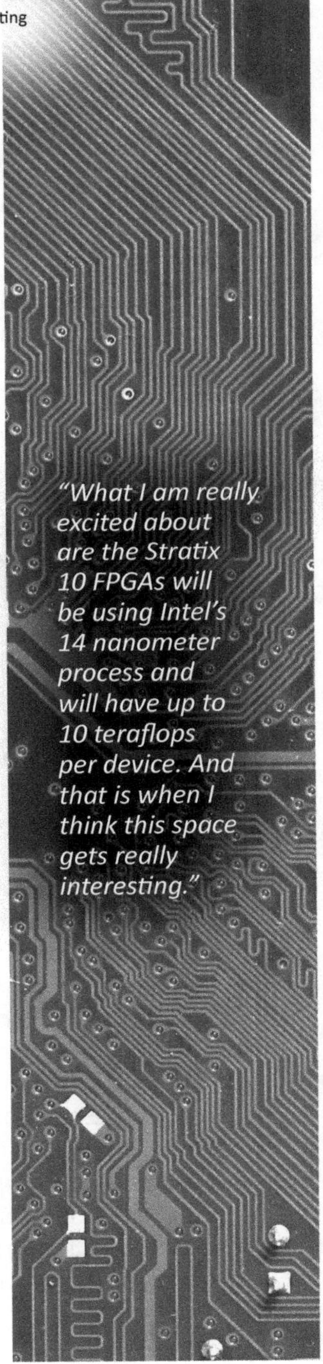
"What I am really excited about are the Stratix 10 FPGAs will be using Intel's 14 nanometer process and will have up to 10 teraflops per device. And that is when I think this space gets really interesting."

demand for a particular task, and when that resource is no longer needed, we can use those resources for other applications. The datacenter has a very diverse set of applications, and this is a very different question than asking what would you do if you just wanted a platform for deep learning. Machine learning practitioners just want the largest GPU cluster than can get, and we are asking a slightly different question here."

The choice of system design involves balancing the desire to have specialized functions for specific applications against the economics and simplicity of having a homogeneous set of infrastructure. Having a relatively homogeneous set of infrastructure allows for components to be purchased at the highest volumes and therefore the lowest prices and also increases the maintainability of the infrastructure because the support matrix on components is smaller. (This is one of the reasons why the X86 server has come to dominate the datacenter.)

The problem arises, however, when you have an application where the work just can't get done fast enough on CPUs, or you have to allocate a tremendous amount of resources to an application to get an answer in a timely fashion. We would say that you end up paying with either time or money, and sometimes both if the job is difficult enough. So it is with deep learning, which only achieved the breakthroughs we have seen when convolutional neural networks with very large training datasets were married with GPUs with lots of relatively inexpensive parallel computing capacity.

"The deep learning community is very happy with this because you have given them machines based on GPUs or ASICs that can speed up the convolutional neural networks by several orders of magnitude," Chung said. "But there are some problems with this. If the demand is low at any given time, you have a stranded capacity issue. Another problem is that you have broken this in the datacenter. If your demand exceeds what you have in your limited pools, you cannot scale beyond that. In fact, within Microsoft we do have a lot of GPUs and the big complaint is that we do not have enough of them. And heterogeneity is incredibly challenging for maintainability. You want to have the minimum number of set of hardware SKUs to test and maintain. So having a very different SKU to fit GPUs is very challenging."

Microsoft could put a GPU accelerators in every single server, which is great for homogeneity and maintainability, but this would substantially increase the power and cost of servers, said Chung. And not all workloads will necessarily make use of those GPUs, even if they could be put into the Open Cloud Servers, which they cannot. (Microsoft could use a 40-watt GPU aimed at laptops and plug it into a mezzanine card much as it does its FPGA, but the GPU is not necessarily as malleable as an FPGA when it comes to the diversity of functions it can perform.) Microsoft could also just create special-purpose ASICs specifically to run neural nets.

As we know from the network appliance business, which uses the bulk of FPGAs sold today, an FPGA is basically a way to simulate the operations of an ASIC. It

"At Microsoft, we have a very diverse range of workloads – enterprise, Bing, Azure workloads, we have latency critical workloads and batch workloads, email – I would be willing to say that the number of servers we needed for running deep learning is at most in the single digits percentage of all workloads. This is a problem for ASICs because there is a long turnaround time to implement them."

is arguably harder to program than a GPU accelerator, but it has its own kind of flexibility in that it can be programmed to do just about anything a chip can do as well as implement the algorithms embodied in software. The downside of an FPGA, at least with the current generations of devices available from Altera and Xilinx, the peak performance of an FPGA is considerably less than for a GPU accelerator. The FPGA therefore represents a balance between something that is general purpose and specialized hardware, said Chung.

What seems clear is that Microsoft wants to have some kind of relative accelerator with a low power envelope inside of its Open Cloud Server nodes. The Stratix V FPGAs from Altera weighed in at around 25 watts, adding about 10 percent to the heat dissipation of the server, while adding only around 30 percent to the cost of the server. Chung was not at liberty to discuss what an Open Cloud Server mode cost and what it was paying for the FPGA accelerator daughter cards in the system. Our guess is the node costs a couple thousand dollars and the FPGA accelerator costs several hundred dollars.

For various kinds of recognition systems, Microsoft uses deep convolutional neural networks, and it wants to use its infrastructure for both training the nets and for running the algorithms that are derived from the training that actually do photo, video, and voice recognition in production applications. (To give you a sense of the scale of these applications, Facebook uses massive banks of GPUs to train neural nets how to analyze photos for their content, and uses CPUs to use the output of these nets to categorize a whopping 600 million photos uploaded by users every day on the social network. How much CPU capacity it takes to do this is a secret, but it is probably not a small amount.)

Microsoft's goal, said Chung, is to get an order of magnitude performance increase on neural nets using the Catapult FPGA add-on for its Open Cloud Servers and fitting in that 30 percent incremental cost and 10 percent incremental power budget. Microsoft wants to expose the FPGA functionality as it relates to deep learning as a composable software library. Incidentally, Microsoft can change the personality of an FPGA in around a second, and has created a different software stack it calls the Azure SmartNIC, which is a network interface card married to an FPGA that can do line rate encryption/decryption and compression/decompression on network traffic as well as other software-defined network functions using the same Catapult mezzanine card.

The Catapult FPGA card has a Stratix V D5 that has 172,600 Adaptive Logic Modules (ALMs) which implement the "code" in the FPGA plus 1,590 hard coded digital signal processors (DSPs) and 2,014 M20K memory blocks (for a capacity of 39 Mbits). The Catapult FPGA mezzanine card has 32 MB of NAND flash memory and 8 GB of DDR3 DRAM memory (two sticks running at 1.33 GHz) for storage and has a single PCI-Express 3.0 x8 link to hook the FPGA to the Open Cloud Server node.

This is very heady stuff, but the cut and dry is that this FPGA block implementing the neural net is statically and dynamically reconfigurable, so you can dial up and down the number of layers and dimensions in the net as well as change the precision of the numbers used in the algorithms without having to rejigger the FPGA's underlying hardware description language.

The Catapult network also allows for neural nets to extend across multiple FPGAs. Chung's colleague at Microsoft Research Technologies, Andrew Putnam, described at last year's Hot Chips shindig. Each FPGA has two Mini-SAS SFF-8088 ports coming off it, which is used to make a private network fabric that is just for the FPGAs and that is separate from the 10 Gb/sec Ethernet switched fabric that links the Xeon server nodes to each other so they can share work and data. The SAS links run at 20 Gb/sec and there is about a 400 nanosecond latency hop across the SAS fabric. Each Open Cloud Server has 48 nodes – 24 half-width servers with two Xeon sockets each – and the torus interconnect works like this: six adjacent FPGA cards are linked in East-West fashion to each other, and then these multiple groups of FPGAs are linked to each other in a North-South fashion to cover all 48 FPGAs in a single rack of Open Cloud Servers. (Microsoft had to create its own six port and eight SAS cables to do this FPGA fabric.)

The point is, this FPGA fabric allows for up to 48 of the devices to be ganged up to work together without involving the CPUs or their network whatsoever. This is one of the critical aspects of what Microsoft has done that makes its use of FPGAs scalable and competitive with GPU or other kinds of accelerators of special ASICs for running neural nets. The FPGA setup has two different modes of operation when running neural net training code, says Chung: one that processes a single image to do a classification as fast as possible and another that works in batch mode that throws a bunch of images at the net and classifies them in parallel.

So how does the FPGA stack up against CPUs and GPUs when it comes to neural net training? Microsoft ran some tests on its Catapult setup and compared it to some raw CPU and CPU-GPU hybrid tests using the ImageNet benchmark to figure that out.

The CPU-only test is for an Open Cloud Server node using to eight-core Xeon E5 processors, and it is running Linux and the Caffe neural net framework. That server node has 270 gigaflops of performance with processors running at 2.1 GHz, and it is able to process 53 images per second on the ImageNet-1K test. (This has 1,000 different image classifications.) Chung says this yields an efficiency of about 27 percent on the CPUs, which is not all that great. The machine had a peak power draw of 225 watts and if you do the math that works out to around 300 million operations per second per joule. Chung did not talk about how the Open Cloud Server equipped with the Stratix V FPGA did, but as previously reported, it can do 134 images per second.

For its most recent deep learning tests, Microsoft upgraded the Stratix V FPGA card in the Catapult system to an Arria 10 GX1150 from Altera, which has a lot more oomph. Microsoft had been projecting it could get about 233 images per second processed on ImageNet-1K using the Arria 10 FPGA, but it has done a bit better than that in its actual tests, hitting 369 images per second. With 265 watts of power draw, that works out to 1.9 gigaoperations per second per joule, which is a lot better than the power efficiency of the plain CPU setup.

That Arria 10 FPGA is rated at 1.36 teraflops of raw performance, and Chung said the FPGA was running at 270 MHz but Microsoft hoped with floorplanning and other tweaks it could boost that to around 400 MHz. With those tweaks, Chung hopes to boost the utilization of the FPGA from about 35 percent of its capacity to 89 percent, and could drive the ImageNet-1K benchmark as high as 880 images per second.

For a GPU comparison, Microsoft is citing the combination of a Titan X GPU graphics card, which has 6.1 teraflops of single-precision floating point math, paired with a Core i7-5930K processor from Intel. With the Titan X doing the heavy lifting, the ImageNet-1K test can burn through 4,129 images per second and that setup burns 475 watts for 11.4 gigaoperations per second per joule.

This brings up an important point. In a deep learning cluster, the CPUs are mostly doing bulk transfers of data to the FPGAs for processing and are mostly idle, says Chung. So if you have a dedicated deep learning cluster, these idle chips consume power but they can't do any real work. In the Azure cloud, Microsoft can put those idle Xeons to work doing something else.

"In this calculation, we are factoring in total server power, but that is not exactly quite right because in the

datacenter we often have many tasks, and the CPUs are mostly idle and you can use them to do other kinds of work. So it is not clear if you really want to include the server," says Chung.

We say this: put the application that uses the neural nets on the CPU part of the Open Cloud Server and have it constantly classifying new photos, then use the FPGA or GPU accelerators that are constantly training the neural nets. This way, you get a frontal lobe, complete.

The point is, the FPGA is not looking too shabby in terms of gigaops per joule, but Chung did point out that Microsoft is still working with underutilized FPGAs.

"We have a small team of developers who have worked on this for a very short time compared to a community that has worked really hard to make this run very fast on a GPU," Chung explained. "We know that we actually have quite a bit of headroom. In fact, these are just projections, and please don't hold me to them, and if we did some floorplanning and scaled out the designs and maximized all of the FPGA, we could push that to around 880 images per second and there we will start to see some very interesting energy efficiency numbers."

Because the Arria 10 FPGAs pack so much more of a wallop compared to the Stratix V FPGAs, Microsoft thinks it can make up the difference between GPU and FPGA training performance with server node scale – particularly with the Catapult FPGA fabric at its disposal. But Chung and his colleagues are already looking ahead. "What I am really excited about are the Stratix 10 FPGAs will be using Intel's 14 nanometer process and will have up to 10 teraflops per device. And that is when I think this space gets really interesting," says Chung.

Microsoft's embrace of programmable chips knowns as FPGAs is well documented. But in a paper released at the end of 2016, the software and cloud company provided a look into how it has fundamentally changed the economics of delivering hardware as a service thanks to these once-specialty pieces of silicon.

When Facebook embarked on its Open Compute Initiative and built its own servers it had a similar goal; to innovate on hardware at a speed closer to software. But in redesigning its very chips for a specific workload Microsoft is taking this idea a step further. Burger looks at this like a different type of computing, likening a general purpose CPU that processes each instruction in an in-order sequence on a processor as temporal computing. FPGA's he says are spatial computing, with an instruction laid out in hardware on the chip and the data funneled through the right path.

According to one source, the Catapult hardware costs less than 30% of all of the other server gear, consumes less than 10% of the power and processes data twice as fast.

Bing was actually one of the first test cases for Project Catapult back in 2014. But after Burger built a server that could work for Bing, Microsoft decided that it didn't make economic sense to have an entire FPGA effort that only worked in one aspect of its business. So Burger

By enabling the FPGAs to generate and consume their own networking packets independent of the hosts, each and every FPGA in the datacenter can reach every other one (at a scale of hundreds of thousands) in a small number of microseconds, without any intervening software. This capability allows hosts to use remote FPGAs for acceleration with low latency, improving the economics of the accelerator deployment, as hosts running services that do not use their local FPGAs can donate them to a global pool and extract value which would otherwise be stranded. Moreover, this design choice essentially turns the distributed FPGA resources into an independent computer in the datacenter, at the same scale as the servers, that physically shares the network wires with software.

started over and built an architecture that could support all of Microsoft's scaled out businesses, from Bing to Azure and even one day, to machine learning.

Because it wasn't enough to just speed up the processing of search on a node, and instead think about how to use FPGAs to speed up things like networking, Burger and his team came up with a different architecture.

Instead of having the FPGAs in a cluster of servers talk to a top of rack switch, the FPGAs sit between the servers' NICs and the Ethernet network switches. Thus, all of the FPGAs are linked together in a network and network traffic is sent through the FPGA network. The FPGA can still be used as a local compute node because it also has an independent PCIe connection to the servers' CPU.

So the CPU can send tasks to FPGA when needed, but the FPGAs can also communicate together to accelerate networking. In this case, the FPGAs can be used as a network processor, which allows Azure to offer incredibly low-latency in its cloud business.

This architecture allows some pretty powerful things to happen. Azure can add or support new networking protocols. Elements such as new encryption technology can be applied universally. And while Burger was cagey when asked about how quickly the FPGAs can be reprogrammed, the sense was that it would take weeks not months.

The biggest challenge for Microsoft as it embarks on this new strategy are poor designs, and trying to apply FPGAs to workloads that aren't big enough to reap the reward. For example, Burger says, "machine learning is not a big enough workload to go to scale yet."

At the end of the day Burger's team has to ensure that every request for a design pays for the hardware cost of designing the new image. Before we had massive workloads that benefited from more specialized and efficient computing FPGAs were an expensive luxury for defense and rarified applications.

Microsoft's real insight is that now FPGAs can be economical for cloud giants. And that it figured out how to use them in a distributed fashion.

* * * *

Well before the Intel acquisition of Altera and the news about Microsoft's use of the Catapult servers, which feature FPGAs to power their Bing search engine and other key applications, it has been clear that the FPGA future is just starting to unfold. With all the pieces in position in terms of the vendor ecosystem, understanding where their value might be for actual applications beyond where one might have expected to find them ten years ago is still something of a challenge. But that picture is getting clearer.

In his 25-year career working with FPGAs, UCLA's Dr. Jason Cong has watched the devices move from purpose-driven implementations, to devices for prototyping and now, in more recent years as computing devices and accelerators. Cong has developed a number of key technologies to push FPGA functionality and programmability forward, something that FPGA maker, Xilinx, has historically been interested in. The company acquired

"Von Neumann architectures were elegant and have served us well, but if you think about something like the human brain, there is no single pipeline to execute instruction after instruction; we have one part for language processing, another for motor control, and so on. These are highly specialized circuits. If you think about Von Neumann architectures as well, executing something simple such as add has many steps—retrieve, decode, rename, schedule, execute and write back—there are usually five to ten steps, depending on what you want to do and this pipeline is not efficient."

one of Cong's developments, the AutoES tool (renamed Vivado HLS after the 2011 purchase and also acquired a scalable FPGA physical design tool via Cong's Neptune Design Automation startup in 2013. Other startups include Aplus Design Technologies, a UCLA spin-out company that developed the first commercially available FPGA architecture evaluation and physical synthesis tool—something that was licensed by most FPGA makers until it was acquired and eventually pulled into Synopsys in 2003.

Much of this development in the 1990s until more recently has been aimed at the prototyping space, but Cong is seeing a new wave of options for FPGAs as compute engines and accelerators for a number of workloads, including more prominently, deep learning and machine learning. "The future will be accelerators with the CPUs there to interface with the software, handle scheduling, and coordination of tasks, but the real heavy lifting will be done by accelerators." This includes GPUs, he says, but the single instruction, multiple data limitation (SIMD) limitations are clear and while FPGAs with their programmable fabric, customizable logic and interconnect, and low power are an appealing option, the programmability needs to be stepped up—and quite significantly.

Although Cong believes that GPUs will continue to dominate on the training side of deep learning, there is great promise ahead for what they might provide on the inference front. For instance, his team put together a study based on the inference for a convolutional neural network and showed the FPGA as capable of offering 350 gigaflops per second in under 25 watts—an attractive story from both a performance and efficiency angle.

Cong did not seem stunned at the $16 billion Intel acquisition of Altera last year, noting the opportunities for efficiency in the datacenter, both for the underlying networking and communications gear that uses FPGAs, but for a new range of application in machine learning, genomics, compression and decompression, cognitive computing, and other areas. "We welcomed this move as well because anything that brings the FPGA closer to the CPU means the latency is reduced—this is something we're working on in a paper that will compare the Intel platform on QPI over existing PCIe. It will be very favorable for many new applications, and even more favorable when Intel puts the FPGA in the same package as they're expected to do in the future." Additionally, one of the startups Cong co-founded, Falcon Computing, which is working with machine learning libraries for FPGAs, they have been able to show how it is possible to reduce machine learning (inference) workloads from four servers to one using such libraries and an FPGA card. Cong says the results of this will be published soon and show a new range of potential machine learning and deep learning capabilities coming to FPGAs.

"There is full confidence in Intel's manufacturing capabilities to bring this all to the same package, but the real risk is on the programming side," Cong says. "Personally, I don't think OpenCL is the answer as it is now. Look at all the datacenters where it's not being used

"FPGAs can be easily integrated into datacenters, adding 20 watts as compute engines in a PCIe slot like Microsoft is doing with Catapult across thousands of nodes. GPUs cannot do that; the CPU already has between 200-300 watts so those will not fit into the server profile. And while you can design an ASIC and it is more efficient than anything else, they are very rigid. This is a time of great, fast change so in the time it takes to fab and ASIC, the algorithms, especially in deep learning and machine learning, could have changed significantly by then."

(big data processing uses Hadoop and Spark, HPC is OpenMP and MPI)." This is something Falcon Computing is focusing on—taking these languages and mapping them to OpenCL automatically without losing performance.

One can imagine how a company like Falcon Computing, which is one of only a few focused on bringing low-level capabilities for FPGA with a higher-level interface might be attractive to the handful of FPGA makers. We know that Cong's previous startups have often found a home at Xilinx, but it is worth questioning what tooling Intel will need to buy into (if at all) to support its FPGA initiatives over the next few years. In other words, we would not be surprised to see Falcon gets snapped up by Intel—or by Xilinx or others at some point in the near future.

Cong's work at UCLA's VLSI Architecture, Synthesis and Technology (VAST) lab garnered him an IEEE Technical Achievement award for work on making FPGAs easier to program. In 2008, Cong and a group of researchers were awarded a $10 million NSF Expeditions in Computing grant to extend over five years, which was based on the recognition then that the days of swift frequency scaling were coming to an end—but he could not have guessed even then that the opportunity for FPGAs as accelerators would be such a large part of the story.

Over the long course of IT history, the burden has been on the software side to keep pace with rapid hardware advances—to exploit new capabilities and boldly go where no benchmarks have gone before. However, as we swiftly ride into a new age where machine learning and deep learning take the place of more static applications and software advances are far faster than chipmakers can tick and tock to, hardware device makers are scrambling.

That problem is profound enough on its own, and is an entirely different architectural dance than general purpose device have ever had to step to. Shrinking dies and increasing reliability and speed are fine arts that have been mastered. But with the new algorithms that rise and take shape, merging and evolving with an incomprehensibly young, rich code ecosystem, chipmakers do not even sure where to begin, let alone how to integrate their ages-old wisdom.

The easy way to address this period of transition is to buy the expertise. The hard way is to build it. And with a machine learning code base that is rich and shape-shifting, there appears yet to be a standard "correct" approach. In the midst of all of this is a fundamental divide between what devices are being dreamed up to suit this new crop of users and what the codes they are deploying actually require, according to Dr. Pedro Domingos, University of Washington computer science professor and author of *The Master Algorithm.*

"In the past there was no real compelling reason to have machine learning chips, but that has changed," Domingos tells *The Next Platform*. Just as machine learning and deep learning were seeing a resurgence, work on the hardware and software side on GPUs was proving itself at scale and, along the way, paving the way

"We need computers that reduce the information overload by extracting the important patterns from masses of data. This poses many deep and fascinating scientific problems: How can a computer decide autonomously which representation is best for target knowledge? How can it tell genuine regularities from chance occurrences? How can pre-existing knowledge be exploited? How can a computer learn with limited computational resources? How can learned results be made understandable by us?"

for other research areas to push into acceleration. And it also just so happened that GPUs were very good at the matrix multiplication-based problems deep learning was chewing on.

That convergence gave Nvidia a clear head start in the marketplace for deep learning training in particular, but as we have been reporting over the last couple of years, others with specialized architectures (from FPGAs, custom ASICs, neuromorphic chips, etc.) have seen an opportunity for catering to the different hardware demands for this segment of the market. Of course, Intel has also see the opportunity, snapping up Nervana Systems and Movidius—both device makers with machine learning optimized software stacks in tow. Despite all of this effort to play catch up and fight battles over who wins the processor shares for the wide variety of machine learning users, there is still a fundamental disconnect from the general purpose processor players and the evolving needs of the diverse machine learning community.

"The big companies right now; Intel and Nvidia, are still trying to figure this space out. It is a different mode of thinking. From a machine learning perspective, we can say what specific primitives are needed from the hardware makers, but the issue is deeper… The machine learning people can tell the hardware people what they want but the hardware people need to tell the machine learning people what they actually can and can't do. It's that interaction that will get interesting results." As it stands now, throwing hardware devices at the wall to see what sticks is a wasteful approach—and even less useful with such a quickly evolving code base.

"Deep learning is just one type of machine learning. Just because there are chips that are good for deep learning doesn't mean they will be good for other types of machine learning," Domingos says. The real question hardware makers need to consider is what will happen when yet a new wave of machine learning comes in; how can the hardware be flexible enough to support it and if there is some set of primitives that can be implemented in hardware, those could change over time as well." There are some primitives that haven't changed much over the last five years during the new golden age of machine learning, and by focusing on these, Domingos says, Intel and other companies can get an early foothold.

"The invention of the microchip was, and still is an amazing thing; it is super-reliable, it's completely deterministic, and it lets us keep building things because solid state electronics are so reliable. No other area has this amazing gift. Not chemical engineering, not mechanical engineering—they have to live with all the crap and noise and things that break. But in computer science, we get to live in the real world; a world where we build programs. And machine learning is taking us back to the real world. It is statistical. It is probabilistic. And we don't know always know why things work or don't all the time."

It can be confusing for us in machine learning but for a company like Intel, where they are used to things being reliable and rigid, machine learning takes a dramatically different mentality. "For machine learning where things are statistical to begin with, not everything needs to be in its place and defined. There is a big opportunity here, but the mental transition to this way of doing things—of letting things work 80% of the time versus near 100% in an application is better than not having anything at all—and better than having a bunch of pre-programmed rules that only catch 5% of the potential use cases."

But the general purpose hardware camps are planting their flags in machine learning now, building the software underbelly as they go; afraid to miss out on a potentially lucrative market. After all, it wasn't long ago that Intel stated that over half of all workloads running in datacenters will have a machine learning component. On a technical level, that sounds overblown—and one has to wonder what Intel means by machine learning. Is this a catch-all definition for advanced analytics or is it actually a new layer of technology layered on top of all of the other database, data management, and other tools and applications as a top-level intelligence tie? If it's not overblown, can we say that in the next five years we will see the death of static analytics? Either way, big companies are taking big steps to plant a stake in the ground—several for different machine learning workloads, as it turns out (Intel's Knights Mill and Nervana/Movidius acquisitions) and Nvidia's many chips for deep learning training and inference (Pascal, M40/M40, Tesla series, etc.).

So with the understanding that general purpose processing options are still lagging behind in terms of real value to the extensive, growing list of machine learning applications, what will win and lose? One answer is to look to custom ASICs, which several of the startups, especially those focused on deep learning, are seeing as the path forward.

"The thing about machine learning that is key is that there are two sides to the problem; the learning and the model that's produced from that learning. Once you've learned a model it's really just a simple program and that is easy to implement in an ASIC. But the problem is, you don't know what that will be until you have the data—so that means a different ASIC for the first learning part." That can be expensive up-front and besides, models evolve, rending an ASIC useless without an ability to rapidly reconfigure—something an FPGA should work well for.

FPGAs are another possible accelerator for deep learning in particular and in fact, neural networks can be seen as a "soft" version of a neural network. The problem here, as with custom ASICs, is that the problem is not known until the data informs it. In other words, all you can get out of an FPGA or neural network is a subnetwork. So while it might do well for part of the workload, it can't do it all.

Of course, GPUs, FPGAs and custom ASICs aren't the only promising hardware trends on the horizon. Domingos points to neuromorphic devices as a promising area to watch. "Building a neuron out of digital devices is already vastly more efficient than generic hardware and software stacks for deep learning. Opinions are divided here but this is one promising path for efficient semiconductor devices that can tackle these workloads well."

For Domingos, the hope is that we will start to see some core primitives baked into hardware and for further unification of the software tooling to support various machine learning workloads. ""We are going to see that unifying frameworks at the software level will migrate into the hardware (logic, graphical models, Bayesian networks, etc.). Things have to become standard for machine learning in key areas, particularly graphs and similarity computations. This is an area where there will be progress," Domingos says, but as he agrees, on the

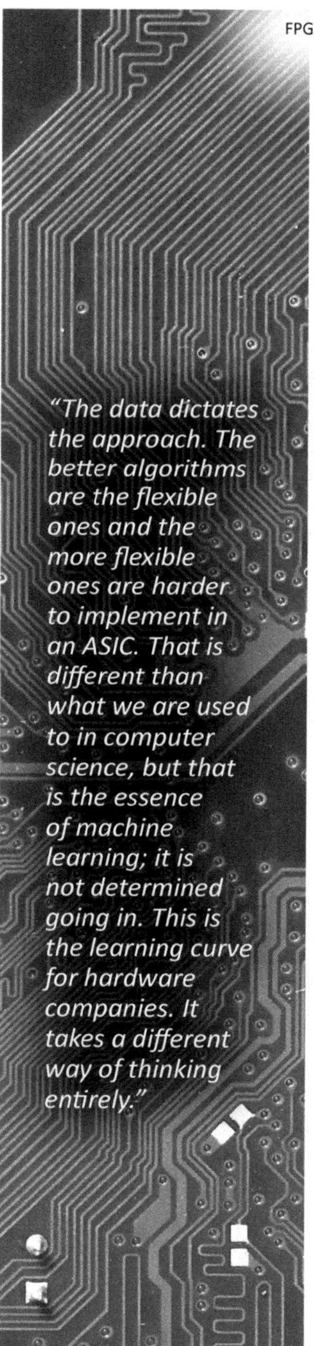

hardware front it's still anyone's game—and a game played on many fields to boot.

Other luminaries see the changing landscape for machine learning and deep learning hardware from a slightly different perspective. Among those we spoke with is one of the so-called "fathers" of deep learning, Yann LeCun.

LeCun is the inventor of convolutional neural networks, which eventually ignited artificial intelligence programs at companies like Google, Facebook, and beyond. Like others who have developed completely new approaches to computing, he has an extensive background in hardware, specifically, chip design, and this recognition of specialization of hardware, movement of data around complex problems, and ultimately, core performance, has proven handy.

LeCun's research work at Bell Labs, which is where his pioneering efforts in deep learning began in earnest, coupled both novel hardware and software co-designs and even today, he is known for looping in the server side of the machine learning and neural network story–something he did skillfully at the 2016 Hot Chips conference. In addition to his presentation on the evolution (hardware and software) of neural nets from his experiences at Bell Labs, Facebook Research, and New York University (among other institutions), LeCun found time to talk with *The Next Platform*about the future of convolutional neural networks in terms of co-design.

Ultimately, he paints an unexpected portrait of what future architectures sit between current deep learning capabilities and the next stage of far smarter, more complex neural nets. What is interesting about LeCun's view is not surprising necessarily: current architectures are not offering enough in terms of performance to stand up to the next crop of deep learning algorithms as it overextends current acceleration tools and other programmatic limitations.

We talked to LeCun about the role of GPU computing for deep learning and walked away with the understanding that the GPU acceleration approach, while useful for training neural networks at massive scale, would have to evolve to tackle the other side of processing–of running the actual algorithms post-training. LeCun extended that argument, explaining that most training models are run on server nodes with four to eight GPUs and that Google and Facebook are working on ways to run training algorithms in parallel across multiple nodes with this setup. He also notes that although the general guess about how many GPUs Google has is somewhere around 8,000, it is actually quite a bit larger than that–scaling as it likely does with growing photo, video, and other datasets.

But perhaps more interesting is the idea that FPGAs are the reconfigurable device that might next on the neural network agenda for processing the larger nets (while GPUs remain the high performance training mechanism). In a very interesting admission, LeCun told *The Next Platform* that Google is rumored to be building custom hardware to run its neural networks and that this hardware is said to be based on FPGAs. Microsoft is testing out the idea of using FPGAs to accelerate its neural nets and is looking forward to the availability of much more powerful programmable logic devices.

The assumption here is that if Google is doing something and Microsoft is experimenting, chances are so is Facebook, along with several other companies that sit on the bleeding edge of neural networks at scale. Although little has helped us come close to understanding the $16.7 billion investment Intel made in purchasing Altera (in terms of astronomically high acquisition sum), statements like these do tend to switch on little lightbulbs. LeCun says that when it comes to the Google and Facebook scale, there is a wariness of using proprietary hardware. "They will actually use their own or use something that is programmable," he noted, which pushes the FPGA door open a slight bit wider.

What began, at the early stages of LeCun's career as a set of tasks to simply classify an image (plane versus car, for instance) has now become so sophisticated that Facebook, one of the most (publicly) extensive users of neural networks for image recognition, can search 800 million people to find a single face in 5 seconds.

The software side of this problem has been tackled elsewhere, most recently in conversations about open source efforts like Torch, Caffee, and other frameworks. But when it comes to the next generation of hardware for both training neural networks and efficiently running them at scale, how is a balance struck, especially considering the relatively "basic" computational needs (fast training on massive datasets, followed by highly parallelizable add/multiply operations?).

These are not new questions for LeCun. During his time at Bell Labs in the 1980s and early 1990s, LeCun and colleagues embarked down an early path toward custom hardware for convolutional neural nets. Among the developments in this area was the ANNA chip, which never found its way into commercial applications for Bell Labs (or elsewhere) but did signify how specialty hardware, in this case a "simple" analog multiplier, could be fine-tuned to chew through neural nets far better than existing general purposes processors. In fact, at the time, the ANNA chip was capable of some impressive feats, including the ability to perform four billion operations per second—quite an accomplishment in 1991, especially for a class of problems that was still emerging.

And if you take a close look at the ANNA chip, it might become apparent that the goals haven't changed much. For instance, the benefit of the chip to circuit design of ANNA is that it could limit traffic with the external memory, which meant the speed of the chip was not limited by how much computation could be put on the processor—rather, it was affected far more by how much bandwidth was available to talk to the outside. That design concept is coming full circle in the multicore world, but what it had in performance it lacked in complexity. And by design, of course. After all, what is the use of having an extensive array of capabilities that aren't needed? And here that "configurability" term raises its head again.

There has already been a fair bit of work done on FPGAs for convolutional neural networks, LeCun says. For instance, he points to an early experiment in scene parsing and labelling in the early 2000s where the team was able to achieve reasonable accuracy at 50 milliseconds per frame using a Virtex-6 FPGA. While this was an ideal processing framework that involved no post-processing of the data, the inherent limitation of Ethernet at the time limited overall system performance. (Similarly, there were other limitations when the next iteration of this idea on Virtex FPGA was rolled out in the NetFlow architecture, which never made it into production due to fab problems—but this was quite a bit later).

Right around the time of the first Virtex-6 work, however, GPU computing was entering the scene, which proved useful for LeCun's continued work. He points to this, coupled with other developments to push image recognition in particular, including the release of the 1.2 million training samples in 1000 categories that were part of the ImageNet dataset as revolutionary new capabilities. The opportunities to train and classify were exponentially increased and the performance of an Nvidia GPU, which at the time was able to process the data at over one trillion operations per second, created an entirely new playing field.

If what is needed to build an ideal hardware platform for deep neural networks is ultra-fast add/multiply capabilities for the actual processing on the neural network algorithm, it stands to reason that something "programmable" (possibly an FPGA) might serve as that base while the the mighty GPU should own the training space. To some extent, it has for some time, with new crops of deep learning use cases from Nvidia and a wealth of examples of large companies that are leveraging GPUs for extensive model training, although not as many for actually handling the processing of the network itself.

While GPUs dominate on the large-scale training side, there are also a number of other emerging technologies that have yet to hit full primetime and await an ecosystem. FPGAs have their place in this expanding ecosystem, but for users at the high end, custom ASICs (despite their cost, long time to market, and lack of flexibility) will continue to be a viable option. Yet another option still will be a custom piece of hardware that captures the benefits of both a custom part and an FPGA—something that is gathering momentum.

The argument is a simple one; deep learning frameworks are not unified, they are constantly evolving, and this is happening far faster than startups can bring chips to market. The answer, at least according to DeePhi, is to look to reconfigurable devices. And so begins the tale of yet another deep learning chip startup, although significantly different in that its using FPGAs as the platform of choice.

DeePhi is a relatively new company (launched March 2016) based on efforts from teams at Stanford and Tsinghua University, As the startup's CEO and co-founder, Song Yao, described at Hot Chips in his introduction of DeePhi (which is short for the phrase "discovering the philosophy behind deep learning computing"), the economics and time to market pressures matched with the rapid evolution of deep learning frameworks make non-FPGA approaches more expensive and less efficient. Yao says that CPUs don't have the energy efficiency, GPUs are great for training but lack "the efficiency in inference", DSP approaches don't have high enough performance and have a high cache miss rate and of course, ASICs are too slow to market—and even when produced, finding a large enough market to justify development cost is difficult.

"FPGA based deep learning accelerators meet most requirements," Yao says. "They have acceptable power and performance, they can support customized architecture and have high on-chip memory bandwidth and are very reliable." The time to market proposition is less restrictive for FPGAs well since they are already produced. It "simply" becomes a challenging of programming them to meet the needs of fast-changing deep learning frameworks. And while ASICs provide a truly targeted path for specific applications, FPGAs provide an equally solid platform for hardware/software co-design.

Overall, DeePhi has developed a complete automation flow of compression, compiling and acceleration which achieves joint optimization between algorithm, software and hardware. A smaller, faster and more efficient deep learning processing unit (DPU) will eventually be released to public. With server workloads on the horizon, DeePhi says they have already been collaborating with leading companies in fields of drone, security surveillance and cloud service. Yao says "The FPGA based DPU platform achieves an order of magnitude higher energy efficiency over GPU on image recognition and speech detection." Deephi believes a joint optimization between algorithm, software and hardware via a co-design approach represents the future of deep learning.

To address the elephant in the room here, if all of these things are true, why aren't we seeing more FPGAs appear in deep learning acceleration conversations? Because, indeed, they're still awful to program.

Yao says that traditionally, FPGA based acceleration via hand coding took a few months; using OpenCL and the related toolchain brings that down to one month. Even still, experimental teams were not getting the performance or energy efficiency numbers they desired. With these facts in mind, the team set about to build workarounds for both the performance and efficiency problems while simultaneously tackling the programmatic complexity.

To these ends, DeePhi has produced two separate FPGA based deep learning accelerator architectures. The first is Aristotle which is aimed at convolutional neural network (CNN) acceleration, the second is called Descartes, which is directed at sparse LSTM (long short term memory) deep learning acceleration. These are matched against work that has been done on model compression and "activation quantization" wherein the team found that getting the precision down (8 bits) is perfectly reasonable—at least on this architecture.

The company's own custom built compiler and architecture is used instead of OpenCL. As Yao says, "the algorithm designer doesn't need to know anything about the underlying hardware. This generates instruction instead of RTL code, which leads to compilation in 60 seconds." This is the result of the hand-coded IP core and design of the architecture as well as the deep compression techniques for model compression. "Deep compression is useful in real-world neural networks and can save a great deal in terms of the number of computations and the bandwidth demands.

Deep compression and data quantization are employed to exploit the redundancy in algorithm and reduce both computational and memory complexity. For both Aristotle and Descartes, evaluated on Xilinx Zynq 7000 and Kintex Ultrascale series FPGA with real world neural networks, the team notes up to 10 times higher energy efficiency can be achieved compared with mobile GPU and desktop GPU. Of course, whether or not this is a fair comparison for server-class deep learning workloads is up to debate since, as we are aware from many interviews, most work at Baidu and other companies is happening on Maxwell TitanX or, at other shops, M40/ M4 combos. Nonetheless, especially for lighter-weight use cases, the results are noteworthy.

In short, while DeePhi might not find its way to critical mass, it shows how researchers are thinking about cobbling together custom solutions to meet targeted workload demands. It's useful to point out that while custom ASIC based approaches have been getting a great deal of attention due to some advantages over general purpose hardware and accelerators, there could be cheaper and, with the right software tools in place, easier options on the horizon.

All of these points about the specialization of workloads and the needs for performance, efficiency, scalability, and cost consciousness lead us to the real point of this book—how will end users consider and adopt FPGAs? In the coming chapter we will take a look at some real-world uses of FPGAs in enterprise, aside from those we've already detailed to highlight trends.

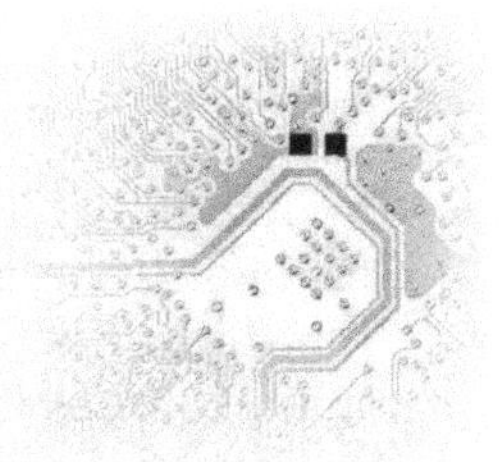

CHAPTER FIVE

FPGA Acceleration in Enterprise

There is a perfect storm developing that is set to whisk the once esoteric field programmable gate array (FPGA) processor ecosystem off the ground and into the stratosphere. Real-time transaction processing, hyper-scale web companies (deep learning/machine learning), large-scale business analytics, signal processing, financial services, IoT, and beyond are all looking at the future of FPGAs for production.

While there were some who have already rolled out platforms featuring FPGAs at the core and backed by a variety of host processors (from ARM to X86), one could suggest they were too early to the game and still required a vast amount of expertise to program (or conversely, had too much sacrifice on the programmability in favor of performance side) and make use of. So what has changed—and with that question in mind, let's use genomics as a prime example.

Aside from the idea that Intel will soon integrate the reconfigurable devices on its chips sometime in the future, there is a growing demand for what FPGAs can lend to data-rich industries, including the genomics market. These reasons go beyond a slowing Moore's Law (in the face of growing data volumes/compute requirements), the added complication of declining costs per genome, and the increase in demand for faster, cheaper sequencing services. Companies like Convey Computing saw this opportunity a couple of years ago in their efforts to integrate FPGAs onto high-test server boards, but following the acquisition of their company by Micron, that future is on hold—and the market is open to new opportunities for FPGA-backed sequencing.

The thing is, FPGAs, even with the OpenCL hooks that companies like Altera and Xilinx have been touting, are not easy to work with, at least to get the ultimate performance. While using a higher level approach like OpenCL will expand the potential market, according to Gavin Stone, who worked with FPGAs for a number of years at Broadcomm, the low-level programming stuff is not going to go away anytime soon. The answer is to take the Convey approach and spin out a company that bakes that low-level programming into the device, match it against powerful host processors (in this case dual 12-core Xeons) and bring that to bear for genomics. The company he works with now, Edico Genome, is a blend of long-time Broadcom FPGA veterans and a slew of genomics PhDs, all of whom are trying to match FPGA and genomics expertise to an ever-expanding market for rapid genomic analysis.

We detailed Edico earlier in this book. The emphasis is not so much due to their products or services being better than others, it is the level of detail we were able to extract in conversations with them and the fact that they were one of the first and only to jump in front of an emerging market with an FPGA-centered offering. This gives them unique insight into how users adopt and use FPGAs and how important abstraction is for end users.

What Edico Genome is doing is interesting on the hardware and software fronts, but the real story of performance is in their claim that they're able to roll out 1000 genomes per day on their in-house cluster, comprised of a combination of Xilinx FPGAs and beefy Xeon CPU cores–all of which combined, comprised their DRAGEN genomics platform. In addition to gunning for the world record there, their server offering is found at a number of institutions, including most recently, The Genome Analysis Centre in the U.K., which is well known for exploring emerging architectures to get the genomics job done faster (including using optical processors, as we highlighted here some months back).

For a team, many of whom came from long engineering careers at Broadcom working with FPGAs, that perfect storm is right overhead—and genomics is at the center. "When we were designing an ASIC to be deployed in production, all the pre-design work is being done in an FPGA, but now, just in the last couple of years, the paradigm has shifted where FPGAs are themselves being targeted to go into production versus as serving as a development vehicle," Stone explains.

The DRAGEN board is a full-width, full-height (similar to GPU form factor) that goes into a dual-processor with two 12-core Xeons, which function as the host processors. There are typically have 6-8 high performance SATA drives in a RAID 0 configuration feeding the DRAGEN to target I/O bottlenecks. The system is strung together with Infiniband and is deployed in a 2U rackmount configuration. The company is selling this as preconfigured servers,

"The general purpose processors are not going to be able to keep up with the data driving all of this, Moore's Law is slowing down or at least becoming a constant for now, and in genomics, which will in the next ten years be the 'biggest of all big data' there is a clear reason why the popularity is growing, and this is why Intel acquired Altera for us, why going into production with an FPGA for production genomics work makes good sense."

which they can scale at will. It is one Xilinx-based DRAGEN FPGA board per host server.

The typical configuration for a high throughput center is several sequencing instruments (Illumina more than likely), which are connected via 10GbE to high performance storage and the data streams off the sequencers to there with a scheduler underpinning it to gather complete runs off the sequencers and over as DRAGEN runs. One DRAGEN card can handle the entire throughput of ten of the highest end Illumina sequencing cluster, Stone says.

There might be two to four DRAGEN servers to support ten or so Illumina setups, but Stone says their implementation in house of 20 DRAGEN servers is the largest to date—and that is the one that is set to garner them a world record for genomes processed per day. While some FPGA approaches have used the host CPU as a secondary element, in this case the 12-core Xeons play a critical role with both FPGA and CPU being constantly fed with utilization Stone says is almost 100%.

Interestingly, Edico Genome has had a partnership to develop on the FPGA with Intel since before the Altera deal was announced. Of even greater interest is that they are currently using Xilinx FPGAs in their custom solutions. And to add one extra meaty bit of information, these genomics machines with Edico Genome's DRAGEN FPGA boards are a bit different than what we've seen in many FPGA installations where the host CPU takes a backseat. Instead of using wimpy ARM or other lower-power cores to feed the system, these systems sport a dual 12-core Intel Xeon processors—letting those CPUs chew on one part of the workload while another parallelizable set of instructions hum away in reconfigurable fashion on the FPGA.

The value proposition for any company using FPGAs as the basis of their workloads is the same, however. To get the most performance out of their applications by blending CPU and FPGA—and exploiting that FPGA to the fullest with programming approaches that indeed, might prove barrier to entry to some, but when done properly, can lend enormous performance advantages. The next big wave of FPGA use will, according to Xilinx and Altera in several past conversations, be driven by the OpenCL programming framework and while this will open a wider user base, Stone says there is still room for specialization, particular in genomics.

"OpenCL makes FPGAs easier to program but it adds an extra layer of abstraction, so it's like programming to Java versus machine code. It's easier for more people, but you don't get quite the performance you would if you programmed down at the machine level," Stone notes. He says that while Intel is a current partner and they will indeed roll out their own FPGA-based servers sometime in the future, this is not a future barrier to their business. "FPGAs are a dedicated processor for the genomics tasks. The barrier to entry is higher but the performance improvements are significantly better once you get down to the machine level with the FPGA and have the domain expertise that you get from hiring PhDs and genomics experts."

"The advantage is being able to run whatever analysis pipelines are needed; that's the advantage of going into production with an FPGA versus an ASIC—there are a lot of different genomic analysis pipelines. We can analyze RNA, agricultural biology, different pipelines for cancer research—all of these have widely varying pipelines and some are just nuanced, but they all require something different to be loaded into the FPGA before that run."

In the meantime, companies like Edico Genome are trying to make the case for why FPGAs make sense for genomics, something that has been a constant push since the company got its start in 2013. There are still relatively few in the genomics industry that truly understand what adding FPGAs into the mix means in terms of performance and capability, but he says that their installation on site where they do genomics runs as a service for existing companies shows that it's possible to take on those 1000 human genomes—something that even large supercomputers can't do. And all of this is done in a relatively small footprint.

Genomics and the life sciences generally has been looking to FPGAs for specific workload requirements, but there are other wider-ranging areas that can make use of FPGAs. Consider large-scale business analytics, for instance.

While much of the work at Chinese search giant Baidu that made news in 2016 focused on deep learning and GPUs, there are some interesting FPGA directions the company has taken for analytics. As Baidu's Jian Ouyang detailed at the 2016 Hot Chips conference, Baidu sits on over an exabyte of data, processes around 100 petabytes per day, updates 10 billion webpages daily, and handles over a petabyte of log updates every 24 hours. These numbers are on par with Google and as one might imagine, it takes a Google-like approach to problem solving at scale to get around potential bottlenecks.

Just as companies like Google are looking for any way possible to beat Moore's Law, Baidu is on the same quest. While the exciting, sexy machine learning work is fascinating, acceleration of the core mission-critical elements of the business is as well—because it has to be. As Ouyang notes, there is a widening gap between the company's need to deliver top-end services based on their data and what CPUs are capable of delivering.

As for Baidu's exascale problems, on the receiving

end of all of this data are a range of frameworks and platforms for data analysis; from the company's massive knowledge graph, multimedia tools, natural language processing frameworks, recommendation engines, and click stream analytics. In short, the big problem of big data is neatly represented here—a diverse array of applications matched with overwhelming data volumes.

When it comes to acceleration for large-scale data analytics at Baidu, there are several challenges. Ouyang says it is difficult to abstract the computing kernels to find a comprehensive approach. "The diversity of big data applications and variable computing types makes this a challenge. It is also difficult to integrate all of this into a distributed system because there are also variable platforms and program models (MapReduce, Spark, streaming, user defined, and so on). Further there is more variance in data types and storage formats."

Despite these barriers, Ouyang says teams looked for the common thread. And as it turns out, that string that ties together many of their data-intensive jobs is good old SQL. "Around 40% of our data analysis jobs are already written in SQL and rewriting others to match it can be done." Further, he says they have the benefit of using existing SQL system that mesh with existing frameworks like Hive, Spark SQL, and Impala. The natural thing to do was to look for SQL acceleration—and Baidu found no better hardware than an FPGA.

These boards, called processing elements (PE on coming slides), automatically handle key SQL functions as they come in. With that said, a disclaimer note here about what we were able to glean from the presentation. Exactly what the FPGA is talking to is a bit of a mystery and so by design. If Baidu is getting the kinds of speedups shown below in their benchmarks, this is competitive information. Still, we will share what was described. At its simplest, the FPGAs are running in the database and when it sees SQL queries coming it kicks into gear.

One thing Ouyang did note about the performance of their accelerator is that their performance could have been higher but they were bandwidth limited with the FPGA. In an evaluation, Baidu setup with a 12-core 2.0 Ghz Intel E26230 X2 sporting 128 GB of memory. The SDA had five processing elements each of which handles core functions (filter, sort, aggregate, join and group by.).

To make the SQL accelerator, Baidu picked apart the TPC-DS benchmark and created special engines, called processing elements, that accelerate the five key functions in that benchmark test. These include filter, sort, aggregate, join, and group by SQL functions. (And no, we are not going to put these in all caps to shout as SQL really does.) The SDA setup employs an offload model, with the accelerator card having multiple processing elements of varying kinds shaped into the FPGA logic, with the type of SQL function and the number per card shaped by the specific workload. As these queries are being performed on Baidu's systems, the data for the queries is pushed to the accelerator card in columnar format (which is blazingly fast for queries) and through a unified SDA API and driver, the SQL work is pushed to the right processing elements and the SQL operations are accelerated.

The SDA architecture uses a data flow model, and functions not supported by the processing elements are pushed back to the database systems and run natively there. More than any other factor, the performance of the SQL accelerator card developed by Baidu is limited by the memory bandwidth of the FPGA card. The accelerator works across clusters of machines, by the way, but the precise mechanism of how data and SQL operations are parsed out to multiple machines was not disclosed by Baidu.We're limited in some of the details Baidu was willing to share but these benchmark results are quite impressive, particularly for Terasort.

On yet another end of the acceleration spectrum is the hybrid role of the FPGA as both server and switch part—something industries like financial services are seeing as boon.

Drawing the lines where the server ends and the network begins is getting more and more difficult. Companies want to add more intelligence to their networks and are distributing processing in their switches to manipulate data before it even gets to the servers to be run through applications. The need for higher bandwidth and lower latency is in effect turning switches (and sometimes network adapters) into servers in their own right.

This is not a new idea, but the server-switch hybrid seems to be taking off in physical switches just as the use of virtual switches are taking off on servers equipped with hypervisors and supporting virtualized workloads. In the financial trading arena, applications generally run on bare metal for performance reasons because nanoseconds count, and so instead of a virtual server-switch hybrid, trading companies want a physical box that has low latency on the switching and enough brains to do useful work that might otherwise be done by servers much further down the line from an exchange.

Because of the competitive advantages that specific technologies yield for big banks, hedge funds, and high frequency traders that invest our money and make a zillion quick pennies on it every second, it is rare indeed to find a financial services customer that will admit to using a particular technology in their infrastructure.

We almost have to infer the utility of hardware or software from the fact that they exist and that companies are selling and supporting them, barring the occasional customer who will talk and explain how and why they chose to implement a particular machine.

So it is with hybrid server-switches, which typically marry high-bandwidth, low-latency Ethernet switches with compute and storage that is sufficient to do a reasonable amount of processing all in the same box. Such a hybrid machine was not just inevitable because X86 processors became embedded inside of switches over the past few years, but that has certainly helped make it easier to move Linux applications that once ran on separate servers or appliances onto the switch itself. Switches have always included field programmable gate arrays, or FPGAs, to help goose the performance of certain functions that are ancillary to the central switching ASIC but which are of too low of a volume to merit being etched directly in silicon. In some cases, switch makers are putting beefier FPGAs in the switches, giving them multiple kinds of compute to chew on data as it moves back and forth across the switch.

This is precisely the tactic that Juniper Networks is taking with its new server-switch, which has a mix of X86 and FPGA compute embedded in it to accelerate financial applications.

Andy Bach, chief architect for Juniper's financial services group, tells *The Next Platform* that the increased message rates required by trading companies, which is growing at somewhere around 30 percent to 50 percent per year, puts enormous pressure on networks. On the CPU front, processor clock speeds are more or less flat-lined at around 3.5 GHz, depending on the architecture, and that means companies can't push performance on single-threaded applications much harder. Aspects of trading applications do not easily parallelize and are dominated by single-threaded performance of CPUs.

And while high frequency trading fostered a market for very low latency Ethernet switches, the switching speed pretty much "bottoming out" at a couple of hundred nanoseconds, says Bach, and so switch makers and financial services firms alike are looking for new ways to boost performance and decrease response times. (Bach ran the networks at NYSE-Euronext for more than two decades, so he has sat on the customer side of the table for a lot longer than the vendor side and knows these issues well.)

But there is, Bach explains, another factor that is pushing companies to rethink their switching infrastructure.

"We are starting to see another step function in processing demand," says Bach. "We went through the era of high frequency traders, and their main advantage was speed and everyone has pretty much caught up. With social media and live news feeds, that really changes the game in terms of what kind of processing capacity you are going to need. Just think about sorting through a live Twitter feed and figuring out what is good news and what is not, what is relevant."

The stock ticker coming off the exchanges–the crawl coming off Wall Street–is barely any data at all by comparison. But the consolidated trade/consolidate quote ticker, or CTCQ, averages around 1 million messages per second, and the options feed can burst to 20 million messages per second and the equities feed is on the same order of magnitude; messages sizes in the financial industry average somewhere between 140 bytes and 200 bytes. A Twitter feed is on the order of millions of messages per second, and other social media sites like LinkedIn and Facebook also have very high message rates that are comparable to other financial feeds. (It depends on if you take the whole feed or just parts of it.)

Live news feeds are also increasingly part of the trading algorithms, and it is funny to think about the news last week that Intel might be acquiring Altera hitting the street actually coursing its way through financial applications, with that very news possibly being chewed on by a mix of Xeon GPUs and Stratix FPGAs and turned into money. CNBC reported that a fast-acting trader (more likely an application, not a person) reacted within seconds of the Altera rumor hitting Twitter and turned options worth $110,350 into a $2.5 million profit in 28 minutes as the market went crazy and drove Altera's stock up 28 percent. That fast reaction on the Altera news could have been a bot waiting for such news to come along, plain old luck, or insider trading.

Suffice it to say, there is still an arms race in the financial services industry, and this time it is about using data analytics to pick the right time to trade, and then executing quickly – not just moving quickly for its own sake.

Juniper would not bother to create a God box switch if there were not customers asking for such a machine, and the wonder is that Juniper has not done it to date with rivals Cisco Systems and Arista Networks already doing it and upstart Pluribus Networks making a lot of noise last year about its own server-switch half-blood. Bach can't reveal the early customers who helped it develop and test is QFX application acceleration switch, but he did tell *The Next Platform* that the Chicago Mercantile Exchange, which runs the world's largest derivatives trading operations, will be using the new switch to route data selectively to its matching engines.

Programming FPGAs is not easy task, so Juniper is partnering with Maxeler Technologies, which has tools that can convert dataflows constructed in the Java language to the Verilog language that encodes the FPGA's functions. The QFX-PFA includes a complete development environment and technical support. Maxeler has expertise not only in financial services, but also in the oil and gas industry and in the life sciences arena and that means this switch could see some action in these markets, too.

The FPGA-enhanced switch from Juniper will have a number of different use cases, according to Bach. Putting the compute in the switch to ingest hundreds of global data sources, combine them, and parse them out to customers with specific feed requirements can be done inside the switches, reducing latency and replacing hundreds of servers. The exchanges have to maintain multiple matching engines for their order books, and this can also be partially implemented on the switches instead of solely on servers. In one test, a customer with a trading plant comprised of 1,000 servers and 1,500 network ports with a 150 microsecond latency was able to use the FPGA to implement part of the work, cutting the server count down to 60 machines and the switch port count down to 1,000 ports while at the same time reducing the latency to under 100 microseconds for the matching engine application.

The exchanges can also use the compute inside of the enhanced QFX5100 switch to do pre-trade risk analysis on the fly, as transactions are coming across the network, with a lot lower latency, assessing if trades are bogus or not. The other kind of risk analysis associated with trading – checking to see if a trade is smart or not – can also be done on the fly inside the switch, says Juniper.

This chapter has only discussed a few of the areas where FPGAs are finding a fit in the enterprise. With further development of machine learning for FPGAs to branch into more analytics, code and other hooks for HPC, and more users tapping into reconfigurable devices in the cloud, this set will grow. With this in mind, we will take a look at how some researchers are thinking about the next generation of FPGA development and how that work can translate into eventual business value.

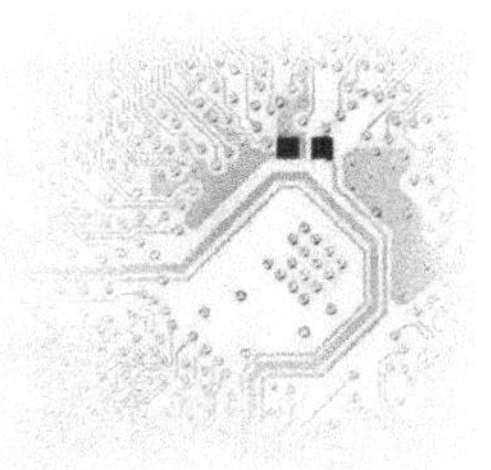

Chapter Six

FPGA Research Horizons

Even if FPGAs do not represent a new technology, the increase in attention from both research and enterprise has been staggering. Accordingly, there have been many research endeavors on both the hardware and software fronts to scrape as much performance as possible from reconfigurable devices.

In this chapter we will take a look at some of the innovative ways researchers are tweaking FPGA devices and programming environment to help foster a clearer picture about what lies ahead. The momentum around this topic in research is similar to what we saw in the mid-2000s with GPUs, which had a few development use cases and then exploded to become an industry of their own.

In the last couple of years, much has been written about the usefulness of GPUs for deep learning training as well as, to a lesser extent, custom ASICs and FPGAs. All of these options have shown performance or efficiency advantages over commodity CPU-only approaches, but programming for all of these is often a challenge.

Programmability hurdles aside, deep learning training on accelerators is standard, but is often limited to a single choice—GPUs or, to a far lesser extent, FPGAs. With this in mind, a research team from the University of California Santa Barbara has proposed a new middleware platform that can combine both of those accelerators under a common programming environment that creates enough abstraction over both devices to allow a convolutional neural network to leverage both with purported ease.

The idea that using programmable devices like FPGAs alongside GPUs in a way that makes anything easier for programmers sounds a bit far-fetched, but according to the research team, which did show impressive results on an Altera DE5 FPGA board along with an Nvidia K40 GPU, the approach can "provide a universal framework with efficient support for diverse applications without increasing the burden of the programmers."

"Compared to GPU acceleration, hardware accelerators like FPGAs and ASICs can achieve at least satisfying performance with lower power consumption. However, both FPGAs and ASICs have relatively limited computing resources, memory, and I/O bandwidths. Therefore, it is challenging to develop complex and massive deep neural networks using hardware accelerators. Up to now, the problem of providing efficient middleware support for different architectures has not been adequately solved."

They also looked at the trade-offs and relative differences between the two devices in terms of energy consumption, throughput, performance density, and other factors, and found it is possible to balance the framework to favor the better device for parts of the workload.

An API sends requests to the scheduling middleware on the host code, which can then offload part of the execution threads to CUDA or OpenCL kernels. These kernels have a shared virtual memory space. The scheduling and runtime support are abstracted from the programmer's view and handled by the API, which serves as the bridge for the applications to either device.

The effort, called CNNLab allows applications to be mapped into the kernels using CUDA and OpenCL using the team's middleware as a bridge between the neural network and the accelerators. The framework is flexible, meaning tasks can be allocated to either device and a common runtime is used for both architectures.

In developing the middleware framework, the team did some interesting benchmarks to understand the relative differences between GPUs and FPGAs for different deep learning approaches using a hardware prototype they built using the Nvidia and Altera parts to understand the differences in execution time, throughput, power, energy cost, and performance density. These results alone are worth a look.

The team's results show that "the GPU has better speedup (100x) and throughput (100x) against FPGAs, but the FPGAs are more power saving (50x) than GPU." They also note that the energy consumption for convolutional neural networks across both devices is approximately similar. In terms of performance density, both are also not far from each other, with the FPGA in the 10 gigaflops per watt range with 14 gigaflops per watt for the GPU. However, they note that the operational efficiency is higher for GPUs.

Although the team was able to show speedups on their platform, they note that there are further developments needed. The speedup of both devices can be enhanced by better compressed network models. Further, they are considering how the accelerators might be paired using Spark or TensorFlow as the data processing backbone.

It is interesting research work in terms of the creation of a bi-directional layer of software that can speak both FPGA and GPU to maximize performance and efficiency, and with some work, one can see how boards based on both with a low-power processor can be strung together for convolutional neural network workloads. but if FPGAs were indeed easy to talk to and the OpenCL/CUDA interfaces were so easy to co-mingle, one has to wonder why this hasn't been attempted already–if, indeed, it hasn't.

The larger story here, beyond the middleware framework the team created, is how the metrics for both accelerators stack up. While Microsoft Catapult and other systems and approaches are using FPGAs for deep learning, it is one side of the accelerator story for this area that has been less touted. We expect that Intel's acquisition of Altera and its now direct focus on machine learning might yield more work in this area in the coming year or two.

Other researchers are seeing a product angle for future FPGA-based systems—something that meshes the familiar with the less ordinary (in this case, the FPGA).

A small company that was founded by Seymour Cray in 1996, just before he died in a tragic car accident, that has been perfecting hybrid CPU-FPGA systems for military and intelligence agencies agrees that there is a future for FPGAs and came out of stealth in 2017 to aim its wares at hyperscale datacenters.

SRC Computers, which is based in Colorado Springs, takes its name from Seymour Roger Cray's initials and was co-founded by the legendary supercomputer designer along with Jim Guzy, one of the co-founders of Intel and until recently chairman of PCI switch chip maker PLX Technology (now part of Avago Technologies), and Jon Huppenthal, who is still SRC's president and CEO.

SRC shipped its first system to Oak Ridge National Laboratory three years later, and the company pivoted away from the HPC market in 2002 when it started selling its hybrid systems to defense and intelligence customers. Except for a few citations of its machines in academic and lab research, SRC has kept a relatively low profile and its 50 employees – all systems, network, and compiler experts who have been around for a while – have focused on making its MAP hybrid processors and Carte programming environment easier to program and deploy than a typical CPU-FPGA setup. Mark Tellez, director of business development at SRC, says that the company has invested more than $100 million in hybrid systems development since its founding two decades ago.

Companies in the financial services and oil and gas industries have tried just about every technology under the sun to goose the performance of their applications and get some competitive edge, and FPGAs have been something they have either toyed with or deployed for specific applications over the years. FPGAs have come and gone as platforms for research in academia and supercomputing labs, too. Supercomputer maker Cray paid $115 million in February 2004 to buy OctigaBay, a maker of supercomputers that married Opteron processors to FPGAs whose products were eventually commercialized as the Cray XD1. And even at SRC, the company backed into FPGAs almost accidentally when it was designing high performance clusters based on commodity CPU chips. The FPGAs were deployed in early SRC systems to replace custom ASICs, and after SRC figured out how to work with them, the company's engineers decided to use them as compute elements in their own right rather than adjuncts.

In a briefing that Huppenthal gave recently, he explained the situation. "We delivered the first system with reconfigurable processors to Oak Ridge back in 1999," he said. "What this showed

us was two things. One, that the use of reconfigurable processors had a lot of merit. And the second thing it showed us is that if you could not program it, it was never going to get used."

As usual, gluing some hardware together into a system is the easy part, but it has taken more than fifteen years and a dozen iterations of platforms for SRC to perfect the application development environment that makes its hybrid CPU-FPGA systems relatively easy to program. "We were able to work through a bit of the issues of taking a sequential language like C and putting it on what is essentially a parallel engine," Huppenthal continued, and he contrasted this a bit to the approach of OpenCL being used as a platform for spreading code out from CPUs to GPU and now FPGA accelerators. OpenCL, Huppenthal said, was a parallel execution environment originally designed for GPUs and it has been modified to run a subset of C on the GPU and now on the FPGA. But there the problem is that you have what Huppenthal calls "an FPGA on a stick," by which he means that the accelerator is hanging off of the PCI-Express peripheral bus, which is too slow, and is not sharing main memory with CPU in the hybrid system. And as such, data has to be moved back and forth between processors and accelerators and something – the CPU – has to be put in charge of the application and decide what gets executed where.

With the MAP hybrid processor and the Carte development environment created by SRC, developers working in either C or Fortran have no idea they are even using a machine that employs FPGAs. The Carte environment has a coding and debugging environment that runs on a client machine. When the code is put into production, it is automagically split between a CPU, which can in theory be an X86, ARM, Power, or any kind of processor a customer needs, and an FPGA. When it started, says Eaton, SRC used FPGAs from Lucent, the HP semiconductor spinoff that is one of the kernels of

Avago. Then SRC used FPGAs from Altera, and when a big generational shift came, Xilinx had the better FPGAs, and now the company feels Altera has the advantage and has been using them. SRC does customized systems – particularly given its defense and intelligence customers – so the MAP architecture and Carte development system has to be fairly agnostic.

The secret sauce in the MAP-Carte stack is actually a system FPGA that implements a shared memory interconnect, called SNAP, between the CPU and user FPGA. This kind of high-speed link for shared memory is something that Nvidia is working to add to future Tesla GPU accelerators using its NVLink interconnect and that IBM has added to its Power-Tesla hybrids through its Coherent Accelerator Processor Interface, or CAPI. SRC figured out how to do this many, many years ago, and because its volumes were not high enough to justify a custom ASIC for this shared memory controller, it implements it on an FPGA along with networking and other functions.

Because of this system FPGA, the Carte development environment sees a single compute resource and a single memory space, and the code is compiled down to the FPGA's hardware description language (HDL) automatically and portions run on the X86 chip in the system as necessary. As Dave Pointer, director of system applications at SRC put it, "instead of running the application *on the processor*, the application *becomes the processor*."

Pointer equates this to the diesel engine coming along and knocking out the steam engine.

The conversion of applications from high level programming languages like C and Fortran down to HDL is the tough bit, and SRC is not giving any secrets away about how its Carte software development tool does this so seamlessly. But with the Carte system and the hybrid CPU-FPGA setup, you get a few things.

The first is that customers can program their hardware to behave any way they want and to have the features – and only the features – they need. If you need 50 floating point units on your system, that is what you put on them. Also, because the application and its dataflows are implemented in HDL and essentially running as hardware, you get deterministic performance. Every time you run a routine, it behaves the same exact way. (You can't say that about a general purpose CPU juggling many things.) Moreover, the FPGA can change its personality on the fly, allowing for mixed workloads on the hybrid nodes over the course of a day. (It takes about a second to change out the application personality on the FPGA, according to Huppenthal – not quite fast enough for context switching speeds inside of applications, but a lot faster than firing up a virtual machine and a whole lot faster than configuring a bare metal server by hand.) Because the FPGA only does the work the application requires (again, the application is the processor), and it executes portions of the code in parallel, the FPGA has a very high utilization rate and, as it turns out, very low power consumption. The combination of high parallelism and low power are what enables a massive amount of potential server consolidation for all kinds of workloads in a datacenter. There are a plethora of use cases for SRC to chase:

This energy efficiency of compute is also, as it turns out, what allows SRC to create signal processing and control systems that can fit inside of a drone, systems that cannot be created given the performance and power draw of traditional, general purpose CPUs. So MAP and Carte are not just a science project, but technology that has been literally battle-tested in the field. And, importantly for financial services, defense, intelligence, and other customers who have FPGA applications already, Carte will allow existing Verilog and HDL code to run on the MAP hybrid processors and be called as routines.

SRC has been building ruggedized rack and mobile servers for the military and intelligence communities for a long time using its MAP hybrids, and it even has a homegrown crossbar interconnect switch called H-Bar that can be used to lash multiple MAP nodes together into clustered systems. But to attack the hyperscale market, SRC decided to partner with Hewlett-Packard and create a MAP cartridge for HP's Moonshot hyperscale system.

The Moonshot machines were launched in April 2013 and started ramping as that year ended. The machines have a very compact architecture and offer a lot of compute density, but given that the "Gemini" Moonshot chassis comes in a 4.3U form factor and has 45 compute nodes, you can't put a very beefy processor on each cartridge, and that limited their appeal in many datacenters. You can, however, put four fairly small processors on a single cartridge, which makes for interesting use cases. Or, as it turns out, you can put on an Intel Atom processor and two Altera Stratix IV FPGAs, as SRC has done to create its Saturn 1 server node:

The interesting bit about the Moonshot machine, explains Eaton, is that the backplane in the chassis has a backplane that implements a 2D torus that has 7.2 Tb/sec of aggregate bandwidth. This interconnect can be used to link all of the nodes in the chassis together (they can be compute or storage nodes) without the need for any additional switching. The two switch modules in the Moonshot chassis are for linking the nodes to the outside world, which can be additional Moonshot enclosures or the upstream network that links to users and other applications. This 2D torus interconnect can link three nodes together in a multiple-tier, as is common for many applications that are based on web, application, and database tiers, or can hook up to fifteen nodes together in a relatively tight cluster that is more akin to the kind used in modeling and simulation applications in the HPC realm. The point is, by switching to the Moonshot enclosure, SRC doesn't have to use its Hi-Bar interconnect to link machines together.

While the Moonshot system is an elegant and clever design, it has not been a barn burner for HP, and part of the reason is that there is limited appeal for the compute elements that can be put on a single cartridge. As to that, Eaton says: "We may be the cartridge that the Moonshot chassis has been waiting for."

The Saturn 1 cartridge has a four-core Intel Xeon Atom processor on the X86 side and two Altera Stratix IV GX530s on the FPGA side. One of the FPGAs implements the SNAP interconnect and multiple virtual Ethernet ports that are used for the 2D torus interconnect and to link to the HP switches in the Moonshot enclosure. In most cases, SRC's early customers are putting a compute node in each row of cartridges and using that as a Linux boot engine from which data gets fed into the remaining 42 MAP hybrid nodes in the enclosure. The MAP nodes do not run an operating system – it is not necessary since the nodes, by design, directly run application code – but customers can boot a Linux kernel on each Atom processor if they want to, according to Eaton.

SRC is charging $19,950 per node for the Saturn 1 cartridge, and that includes the bootloader code required by Intel and Altera for their respective Atom and Stratix processors as well as a license to the Carte environment. Volume discounts obviously apply. But given the kind of consolidation that SRC is projecting for its workloads, customers might not be buying very many of them.

In his presentation, Huppenthal ran some text search benchmarks on a Saturn 1 node and a two-socket server using a high-end quad-core Xeon W3565 workstation processor. (This one has a 3.2 GHz clock speed.) Based on the benchmark results, it would take 1,276 racks of these two-socket servers to do the same work as a rack of Moonshot systems equipped with nine enclosures and a total of 378 MAP hybrid processors. That is a total of 51,040 server nodes and 408,320 cores on the Xeon side and over 10 megawatts of power. The kind of thing you might see at the National Security Agency, for instance.

This comparison is a bit unfair, since those Xeon W3565 processors date from 2009, back in the "Nehalem" days. The compression is not really 1,276 to 1. (And SRC knew that when it made the comparison, obviously.) If you built a cluster using two-socket machines based on the latest "Haswell-EP" Xeon E5-2667 processors, which have eight cores running at 3.2 GHz and considerably more instructions per clock than the Nehalem Xeons did, you would need about 314,406 cores to do the same work, and if you do the math, that would be 19,650 nodes. If you had dense pack, four-node hyperscale machines, that would work out to 234 racks of X86 iron, not 1,276 racks. That is still a huge number of cores, it would still be several megawatts of power, and it is still a huge compression ratio for this text search example if the Saturn 1 performs as expected.

In general, what SRC is saying is that across a wide range of applications that can be accelerated by FPGAs, customers can expect to get around 100X performance with 1 percent of the energy consumption, 1 percent of the footprint, and at under 25 percent of the cost of equivalent performance for X86 server clusters.

SRC trotted out one of the early customers for the Saturn 1 hybrid server node, an advertising startup called Jingit that is based in Minneapolis, Minnesota and that is prepping its first product to come to market in 2017. The company was, as most hyperscalers do, building its stack on X86 clusters, but the nature of the service it is providing made Jingit take a stab at using the SRC system. (SRC has a development lab in the area, so somebody clearly knows somebody here.)

What Jingit wants to do is provide consumer marketing services at the point of sale, which in plain English means kicking out a custom coupon to you as you are buying something. The difficulty is that their backend system has to come up with whatever deal is appropriate for you, based on this current purchase and past purchases, in somewhere around 50 milliseconds to 130 milliseconds, which is the time it takes for a credit card authorization to occur. It was taking a lot of iron to make this happen on the X86 architecture, which was bad, but the total processing time was the real problem.

Including the inbound and outbound message time on the credit card networks, the processing time to be at the current industry benchmark rate should be around 50 milliseconds, with the spare time to compute anything being a very small fraction of that – single digit milliseconds. But the processing time on the X86 clusters was pushing the total transaction time to around 1,500 milliseconds. And 1.5 seconds in our impatient world is a lifetime, apparently. Running the same application on the Saturn 1 nodes, the calculations were being done in nanoseconds, and the response time was quicker than the jitter in the credit card network. Todd Rooke, one of the co-founders of Jingit, says that it is not possible to measure the speed it is happening so fast and the speed of the credit card networks is so variable.

SRC has been attacking various markets where you would expect to see FPGA acceleration for many years, but the difference this time around is that with the combination of the Moonshot chassis and its more sophisticated Carte programming environment, the whole shebang is perhaps a little easier for prospective customers to consume.

While SRC has the backing and history to launch into this space with the intelligence community and others, approaches from other companies taking a hybrid FPGA and CPU slant are also becoming more common.

For some users in high performance computing and a growing range of deep learning and data-intensive application segments, the shift toward mixed (heterogeneous) architectures is becoming more common.

This has been the case with supercomputers, which

over the last five years have shifted from a CPU-based model to a growing set of large systems outfitted with GPUs to add number crunching capabilities. But as both the hardware and software ecosystems around other accelerators gather steam, particularly on the FPGA front, there is momentum around building systems that incorporate a range of accelerators in the same system, which handle different parts of the application.

As one might imagine, none of this is simple from a programmatic point of view—the FPGA piece alone is riddled with complexity landmines. This opens the door for approaches that find a way to automatically discover and assign the right processor—be it GPU, FPGA, or the good old fashioned CPU—preferably at runtime. This might sound like a mythical software layer, one that can "automagically" mesh these multi-accelerator approaches together and auto-select the right tool for the right job, but one HPC startup says it has stitched together such a layer and can provide it as a software-only addition to the heterogeneous system stack, as an appliance, and via a cloud pulled together on Rackspace (and other providers) where FPGAs are offered and can be used alongside a CPU application—and in the same way from the application's point of view.

And for a startup rolling out of the supercomputing world, the fact that they've secured funding to do this is in itself noteworthy. It is (unfortunately) quite rare that true high performance computing startups emerge with significant funding, but Bitfusion, the company with this multi-accelerator approach, sees a path ahead that will extend beyond the hallowed halls of supercomputing and find a fit for the areas where GPU and FPGA-based systems will thrive.

The key, according to the company's CEO, ex-Intel product developer, Subbu Rama, is to be able to do all of this meshing seamlessly—which means that code changes and other complexity barriers need to be removed. After all, he agrees, FPGAs are notoriously difficult to program, a fact that limits their market, even with all of the progress that's been made with OpenCL by companies like Altera and Xilinx. The goal of BitFusion is to provide this software layer and ensure that no code changes happen, even with such smart doling out of accelerator and CPU resources happening at runtime.

"When you have a system that has different architectures, including these three processors, our software layer automatically discovers what device is available, what part of the application is right for what device, and then it automatically offloads that part to the right device—again, without code changes.

Specifically, the code work they've done sits on top of OpenCL to talk to the various processors, which works with Nvidia GPUs and both Altera and Xilinix FPGAs. "We believe OpenCL will be the basis for multiple platforms, but it is not enough. Generally, if you have an application that you want to move from CPU to run on an FPGA now, there is a lot involved. We are building the software libraries in OpenCL, which are performance portable and highly parameterized, meaning it's possible to run those same libraries on any of those three processor types. We discover what hardware is available and dynamically swap our libraries and that's what makes us different than other approaches that have tried to do similar things in the past."

Although this may sound like it would have to add some overhead, Rama says that if one looks at the majority of applications, they are using a finite set of fundamental building blocks and libraries. What BitFusion has done is to take those swaths of open source libraries and moves them into a common OpenCL framework where they can be performance portable across CPU, GPU, and FPGA in a way that is transparent to the user. Internal benchmarks in a few key areas Rama and his team think will be viable markets for the young company include bioinformatics (where they reported an 10x speedup with Smith Waterman), scientific computing (they claim an 11x improvement for NNLS), and data analytics (based on R, NumPy, and Octave for a 31x boost).

There are three ways to make use of the company's work to tie together these multiple processing types. First, as a standalone software layer, called BitFusion Boost (BT), which can be used as either an agent that can sit on top of the OS or as an accelerated Linux image users install on their machines. In either case, the pricing can be shifted to either a per node basis or with a license fee. While we were not able to glean the licensing fees, Rama says BT is currently priced somewhere between $1000 and $2000 per node, according to Rama, a range which is variable based on how they are getting an early handle on user adoption and how the first sets of use cases take shape. One can imagine the market for this is going to be rather limited since only a certain set of users will have machines outfitted with both FPGA and GPU accelerators.

In addition to this software-only model, the company is making an appliance featuring a low-end Xeon, which Rama says is used mostly for orchestration. Configurations can include 4-6 FPGAs from either Altera or Xilnix and on the GPU front, the system can support either an Nvidia K40 or K80, but what is most interesting to Bitfusion is the Titan for its high performance and far lower price point. These come with Bitfusion's software pre-installed in 1U, 2U and 4U form factors. Further, as noted previously, the company is also working with a few key cloud providers on what it calls the BF Supercloud, which lets users tap into an FPGA-enabled cloud provider's boxes for automated acceleration.

Although this might sound compelling for some early stage users who have already seen the performance of the trio of accelerators in action, Bitfusion does not want to be in the hardware business. "We may end up going away from the appliance model at some point,"

Rama explains. "We are experimenting with the model and cost and eventually we won't want to be selling hardware, we would rather channel through OEMs and sell our software, which is the key piece."

Aside from the potential range of applications and use cases this opens for users with existing FPGA and GPU combination systems (or those who might consider an appliance that is already equipped with such a setup) the fact is, this might be a leap forward in terms of expanding the FPGA in particular into new arenas. Rama says the use cases for FPGAs in particular are expanding, noting that what Bitfusion is doing by making it seamless to move from the CPU to the FPGA, is allowing companies to cut potential costs of FPGA experimentation. "Users are not going to be moving everything to the GPU, and certainly not to the FPGA. It might be 10% of an application that they are moving to a new device. They want to see how it works, how it performs, and that can assist with making a decision about whether or not to use a device."

Of course, all the hardware tweaks are lost without ample research work on the software side.

Systems built from commodity hardware such as servers, desktops and laptops often contain so-called general-purpose processors (CPUs)—processors that specialize in doing many different things reasonably well. This is driven by the fact that users often perform various types of computations; the processor is expected to run an Operating System, browse the internet and even run video games.

Because general-purpose processors target such a broad set of applications, they require having hardware that supports all such application areas. Since hardware occupies silicon area, there is a limit to how many of these processor "cores" that can be placed—typically between 4 and 12 cores are placed in a single processor.

If the user develops an application that contains a lot

of parallelism, then that application is still performance-bound by the number of processor cores in the system. Even if an application can utilize 100s of processors, in practice, the application will only see no more than 12 cores.

A new trend is that general-purpose processors can utilize data parallelism with vector instructions, the latest version of intel AVX can process operations on 512 bits in parallel, and the use of GPUs for data-parallel programs. This, however, introduces a second level of parallelism in addition to the cores which may be difficult to utilize in many applications.

A team from Sweden's Royal Institute of Technology (KTH) sought to overcome these limitations of general-purpose processors and data-parallel vector units/GPUs, using Field-Programmable Gate-Arrays (FPGAs). FPGAs are devices that contains logic that is reprogrammable, which means it is possible to create a system inside the FPGA that is tailored to the application; in other words, there is no support needed to run an OS – it only requires the functionality to run a specific application as fast as possible.

To better understand how the Swedish team was able to create a high-level synthesis flow that can automatically generate parallel hardware from unmodified OpenMP programs, we had a chat with the authors of the full paper, Artur Podobas and Mats Brorsson.

Where is OpenMP lacking and how are FPGAs central to making up for the lack of parallelism you point out?

KTH Team: By understanding the needs of the application we can create a system containing a large amount of very small "cores" on our FPGA and be capable of executing many more things in parallel than a general purpose processor ever could; even if the clock-frequency is lower on the FPGA compared to the CPU, the performance benefits of using more parallelism is often worthwhile.

While OpenMP as of version 4.0 contains support for accelerators (through the #pragma omp target directive), so far, the supported accelerators in compilers have been limited to either GPGPUs or Intel Xeon PHIs. Our belief is that OpenMP can be used to drive parallel hardware generation. Most of the existing OpenMP infrastructure is general enough to allow this.

There are a few things that OpenMP could have to better support High-Level Synthesis. Among these are ways for the user to specify other – non-IEEE compliant – floating point representations.

How exactly are you generating parallel hardware from OpenMP programs?

KTH Team: We created a prototype tool-chain that automatically transforms OpenMP applications into specialized parallel hardware.

The hardware that we automatically generate consists of a master soft-core processor and a number of slaves called "OpenMP accelerators". The master processor is thus general-purpose and is responsible for exposing parallelism and schedule them onto the OpenMP-accelerators, where they are executed.

Each OpenMP accelerator contains a number of what we call "hyper-tasks". A hyper-task is a processing element that specializes in performing a certain computation. In fact, the hyper-task is created such that it is only capable of executing a certain computation. Unlike general purpose processors, the hyper-task will only contain logic that is crucial for the computation – it will be void of functionality such as TLBs, large physical registers or out-of-order logic. Instead, we decide at compile time how many registers a hyper-task will have and statically schedule the instructions onto a finite state machine. The real-world analogy here would that if you have a stop in your sewer pipes you would call a plumber and not a baker – the plumber is specialized at fixing the sewer pipes and do so most effectively.

Each OpenMP accelerator contains many such hyper-tasks, which can share certain hardware units between them. The ability to share resource is crucial as many resources are very area-expensive to implement on the FPGA. For example, a floating pointer adder is a costly resource that – if sparsely in the application – should be shared across as many hyper-tasks as possible. The real-world analogy would be that most people own their own mug at work while sharing the coffee machine.

Most parameters such as the number of hyper-tasks per accelerator, the number of accelerators and the internal mapping of instructions to hardware is solved using constraint programming approach, which means that we fully model the problem and time alone dictates how good the solution will be – our methodology generates better hardware with future systems.

Initially, the user provides our prototype tool-chain with an OpenMP application written in the C programming language (we do not support C++). Our tool will parse and understand this application and divide it into two categories: software-based and hardware-based. More specifically, we divide everything related to parallel computation (more specifically, OpenMP tasks) into hardware-based and everything related to exposure and management of parallelism in software-based.

Following the said classification, the hardware-generation step begins. We extract the computational part of the application and transform them into a less-abstract format (the so-called intermediate representation) and perform some standard optimizations on it.

Once the computation parts have been optimized, we create a constraint programming model that solves all variables related to the hardware generation. This includes how each instruction is mapped to the hardware hyper-task, how many of each resources we need and whether they are to be shared inside an accelerator or not. We also solve how many hyper-tasks each accelerator contains.

When all parameters have been solved we generate the entire system. This primarily glues together all accelerators through a shared interconnect (Altera Avalon) together with various peripherals and a general purpose processor. We also create a unique memory map for the System-on-Chip, which states where in the memory each device is located. A memory-map is essentially a map of addresses where the different peripherals and accelerators can be found. This memory-map is used in the next phase: source-to-source compilation.

Source-to-source compilation changes all OpenMP-specific directives into using the underlying OpenMP runtime-system. In our case, it is crucial that this phase comes *after* the hardware has been generated, because both the runtime-system and parts of the translated application requires knowledge about the hardware (e.g. number of accelerators, memory map etc.). The output is a translated C source code that is to be used with the general-purpose processor in the system.

Finally, the application source code is compiled and the generated hardware is synthesized, technology mapped and place and routed to yield a bit-stream. This bit-stream is used to program the FPGA and the user can start using the enhanced OpenMP-accelerated application.

In your conclusions section you say that a main limitation is that you're not modeling the memory hierarchy. Please explain this limitation and what you might do in the future.

KTH Team: One of the main limitations of our work is that we do not take memory hierarchies into account when generating hardware. This includes everything starting from the hyper-task's data request until it is satisfied. In our current work, the OpenMP accelerators do not have a cache and instead rely on hiding as-much of the memory latency as possible by executing concurrently.

One of the biggest challenges in encouraging people to use FPGAs is to reduce complexity. While adding a simple data-cache would have improved the performance of the accelerators, currently it would also have required the user to specify the details regarding the cache, which is what we are trying to avoid. Ideally, everything should be automatically derived and modelled, including caches and interconnect width. Today, we do not model the memory.

Future work of ours focus on analyzing memory traces generated by the application offline in order to satisfy the bandwidth and latency requirements of the targeted application.

Whom do you expect to be most interested in this? Where do you think developments in this will eventually land and what needs to happen first?

KTH Team: We believe that the traditionally embedded High-Level Synthesis marked is slowly moving towards the more general-purpose and even HPC communities. Field-Programmable Gate-Array technology *can* be more power-efficient with respect to CPUs or GPUs. The main focus should right now be to raise the level of abstraction of using FPGA for users.

Thus, the first prerequisite for mainstreaming FPGAs for general-purpose or high-performance computing will be too improve productivity through the use of high-level programming models such as OpenMP but also though the use of data-parallel languages mainly used for GPUs such as OpenCL. CUDA is not likely a candidate as it is too tied the Nvidia hardware.

The second aspect is performance. FPGAs can outperform general purpose CPU but have difficulty in beating GPUs. That is mainly because we are attempting to replicate the GPUs functionality within the FPGAs: we use similar memory hierarchy, similar techniques for hiding latency and – perhaps more importantly – we use the same representations for precision.

One opportunity is to move away from the IEEE standard mantissa/fraction settings for floating point precision, if the application does not require the full precision of the IEEE format. For example, if a user application requires a third of the precision provided by single precision, generating hardware with lesser precision will not only increase the overall peak performance of the system but also benefit data transfers (more numbers transmitted in the same time) and lower power consumption.

Even if we continue to use the IEEE floating point standard, then we can still improve performance by making the FPGA slightly more coarse-grained and introduce hard IP blocks dedicated for floating point performance. This would also increase the peak performance delivered by future FPGA and seem to be the route manufacturers are taking (e.g. Altera Stratix 10).

Leaving off the KTH team's thoughts, in our own analysis of research for FPGAs in 2016, a larger wealth of new published results can be found on the software tuning/programming side for both the device stacks (compilers and tools) and application-specific programming than on the hardware itself. Efforts like those we interviewed KTH about are common as centers try to integrate FPGAs without all of the down-to-silicon expertise. We expect that in 2017 the trend of adding ever-higher level interfaces and hooks for common programming environments will continue.

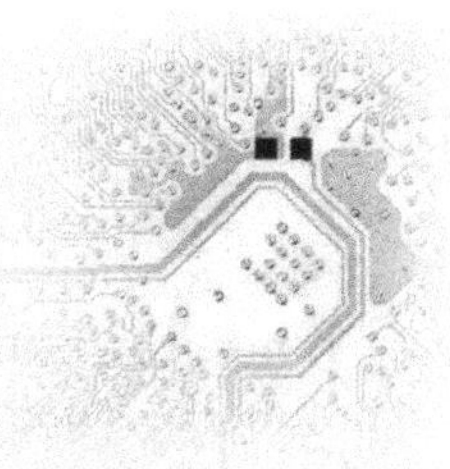

CHAPTER SEVEN

The Future FPGA Market

Now is the time where we pull together all of the things we have examined in this overview in terms of hardware, software, and end user developments for a broader look at the systems and device market for 2017 and beyond

For those who marveled at the $16.7 billion deal Intel made to acquire field programmable gate array maker, Altera, an equal number raised eyebrows at the estimate given by Intel CEO to announce the purchase that one-third of cloud workloads would take advantage of FPGA acceleration by 2020.

It is worthwhile to go back to this stunning assertion about the future applicability of FPGA technology at such grand scale with this in mind. For FPGA silicon that has found its sweet spot, until the last few years in particular, inside specialized datacenters for financial services, oil and gas, and of course, defense and embedded applications, this sudden rise to fame is striking. After all, it was not that long ago that Altera noted its total addressable market in the datacenter was somewhere in the $1 billion range.

So what does Intel see that completes the puzzle—that makes this hefty acquisition make sense from an addressable market perspective? We were able to get Altera to talk about how the market is shaping up for FPGAs, as well as how their perceptions about the total growth potential has shifted over time, especially when the chatter about programmable logic devices hit fever pitch just over a year ago when Microsoft made bold predictions about how FPGAs might power more applications beyond its Bing search and image recognition operations.

As Microsoft's Technology and Research Group vice president, Harry Shum, told a crowd at the Ignite conference in May 2016, Microsoft wants to be ahead of the Moore's Law curve before it rounds out and leaves it scrambling for more ways to deliver key services. This is not unlike how the other hyperscale companies are thinking either, with both Amazon and Facebook and others keep open minds about building increasingly heterogeneous machines.

Following success accelerating the Bing page ranking algorithms, his teams started to look to other services that could be similarly pushed with FPGAs, including machine learning and deep neural networks. "The aspiration," Shum explained, "is that we will build this new fabric of programmable hardware to complement existing programmable software frameworks. And we will build that hardware to benefit a lot of our workloads, then open it up for third party and our own ecosystem as well."

And so, bingo. No pun intended. It is this model of building new datacenters around software, which is of course, designed to be programmable—with hardware that is also programmable to expand a new range of services that are literally built for these devices. And it does not end with Microsoft. Presumably, Intel sees a big opportunity for FPGAs in cloud datacenters (even if a company like Google still sees them as too difficult to integrate into their workflows—and GPUs too, for that matter) if it can make sure to wrap a software ecosystem around them—something that some clever companies (including that small company founded by Seymour Cray that has done some very interesting work programmatically speaking) are doing beyond the OpenCL approaches vetted by Xilinx, Altera, and others.

Altera's head of strategic markets, Mike Strickland, spoke with *The Next Platform* following the Intel deal and while of course he is on lockdown detail-wise (and much of the commentary from this point lies in what Intel will do with its newfound FPGA glory), he was able to offer some clarification about why companies like Intel are seeing a big future in what was once considered a "limited" marketplace (to the $1 billion point, anyway).

This is all a bit cryptic, but one can assume that FPGAs will not only be used for user-facing services, but for users to access, particularly in integrated form on its Azure cloud. And if that happens, it won't be long before (if it doesn't happen before) Amazon offers FPGA cloud offerings for key workloads. That is definitely not out of the range of possibility, either, since AWS was the first large public cloud provider to provide GPU computing instances and tend to stay ahead of the curve by offering the latest cloud-tuned high-end Xeons. This could also mean that companies like Microsoft, who have well-developed compilers and tools (think about OpenCL plus Visual

Studio) can find ways to add meat to FPGA-based servers. Further, Microsoft has done a great job in showing some interesting systems (with accessible OpenCompute designs that Hewlett-Packard, Dell, Quanta, and others can build) that show how to network FPGAs to talk to one another over SAS interconnects while the rest of the network handles other work, creating what is essentially two machines in one—all of which Microsoft released as a production concept last year. The point is, there are endless ways one could look at the potential for FPGA-based datacenters and for the first time in a long time, we're looking to Microsoft to show what might be next in the datacenter from both a hardware and software perspective.

This part is just informed speculation, of course. But with that massive of a cash deal, we have to believe that Intel knows something about the future of the datacenter. It's no secret that most chip and system vendors see a more heterogeneous future ahead—but it would not surprise your friends here at *The Next Platform* if at some point, Altera rival, Xilinx, gets snapped into the maws of another giant. And for purveyors of software that can play well with FPGA-laden systems, the time to market explosion, as much as can happen in this niche anyway, is probably right about now.

Speaking of the code, hooks for FPGAs (which as we've described previously, have a programming hurdle to cross), there may be some momentum there as well. Strickland pointed to the story of GPUs as a solid reference point on how a niche accelerator can build a robust software ecosystem around it, dramatically expanding its reach by reducing some of the complexity. And Altera is standing by OpenCL as the path to swim further into the mainstream.

"If you looked at FPGAs five years ago, Intel wouldn't be interested if you had to do HDL programming for each use and customer. So yes, something changed in that time. One of the big breakthroughs, the biggest building block, is indeed OpenCL. It's not everybody though, but it has come a long way." Strickland continues, "If you look at what Nvidia did, they realized CUDA was too low-level so they advocated OpenACC. There's no reason why something similar can't be done as an extension of the work we've done on our OpenCL compiler. That has a front end that parses OpenCL but most of the heavy lifting is done at the backend of the compiler. There it does over 200 optimizations (external memory bandwidth, for example) and does things that would take an HDL programmer six months to do." With that in place, he says that there is no reason why it's not feasible to add higher-level front ends on the compiler, including OpenACC, OpenMP, and more.

The data and external memory interface management have come a long way over the last few years, Strickland says, and these are the core improvements that moved FPGAs and their OpenCL frame over the accessibility border—well past the type of low-level HDL work that gave FPGAs a bad rap usability-wise for a number of years

While Strickland would not explain in more detail how we went from the $1 billion total addressable market number in the datacenter to one third of all cloud datacenters by 2020 figure—and the mathematics on the Intel side continue to make eyes spin, there is no doubt that this is, as predicted before any of this blew up, the year of the FPGA.

According to the newest IEEE device roadmap, which seeks what lies beyond Moore's Law architecture-wise for a wide range of critical workloads, others are considering similar questions. For some organizations, including the Global Semiconductor Alliance (GSA) and its member organizations, one answer to the post-Moore's Law conundrum lies in pushing the future of open source hardware. While this can come with its own

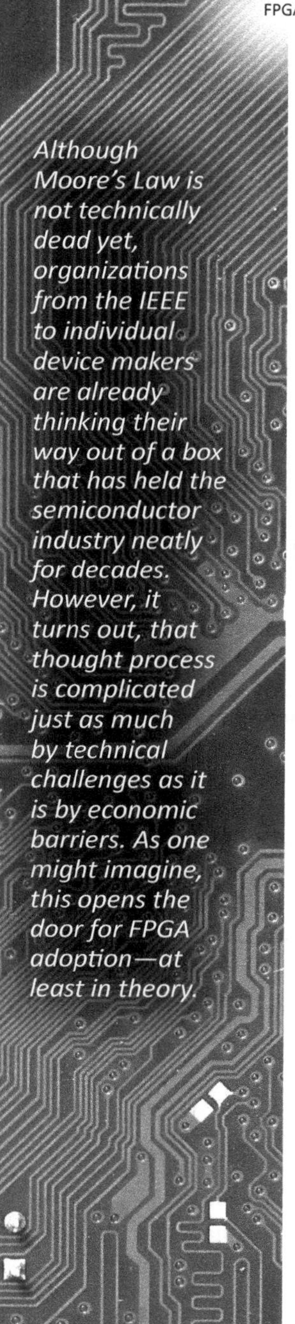

technical hurdles when compared to what the very few large chip-makers produce, the economics, once followed with the right support models to make such architectures viable in enterprise, will follow suit.

Let's bring all of this back to those broader trends for compute (and acceleration) we have touched on throughout this book.

As Jerome Nadel, whose company, Rambus, sponsored a GSA recent report that looks to the future of the industry[1] with open source hardware at the base (as well as other approaches, including reprogrammable devices like FPGAs and custom ASICs) says, "The semiconductor industry is not growing; there has been unparalleled consolidation and money spent on acquisitions recently, and all of this is coming from the fact that this is a non-growth market. The industry is only reaping 1.5 percent of the billions in value it creates, so what we are asking is what alternative paths exist."

Nadel, who works for GSA member company, Rambus, thinks that one potential route is to push the future of open source hardware. Hewlett-Packard Enterprise, Oracle, and others see value in RISC-V, but of course, there are other approaches that are trying to get a foothold, especially in the server market. The device and IoT market are awash in open source hardware options already, but for new devices to become adopted, this means a support angle needs to be firmed up, similar to what Red Hat did with Linux—a move that made Linux overall a more viable option in the enterprise datacenter.

Worldwide chip sales increased by nearly 50 percent during the last decade, with industry revenue peaking at $340 billion in 2014. While those might sound like dramatic growth figures, the numbers mask an even more important story that really started to play out last year. Gartner figures estimate that the worldwide chip sales went down 1.9 percent in 2015, and the 2016 estimate shows only a slight, optimistic 1.4 percent increase. Further complicating the market is the increase in acquisitions and consolidations, driven by the fact that only a few can compete given extraordinarily high design and fabrication costs.

According to the GSA, "Chip design projects that once cost a few tens of millions of dollars a decade ago have climbed as much as $200 million. Growth in the number of IP blocks, use cases, and fuse

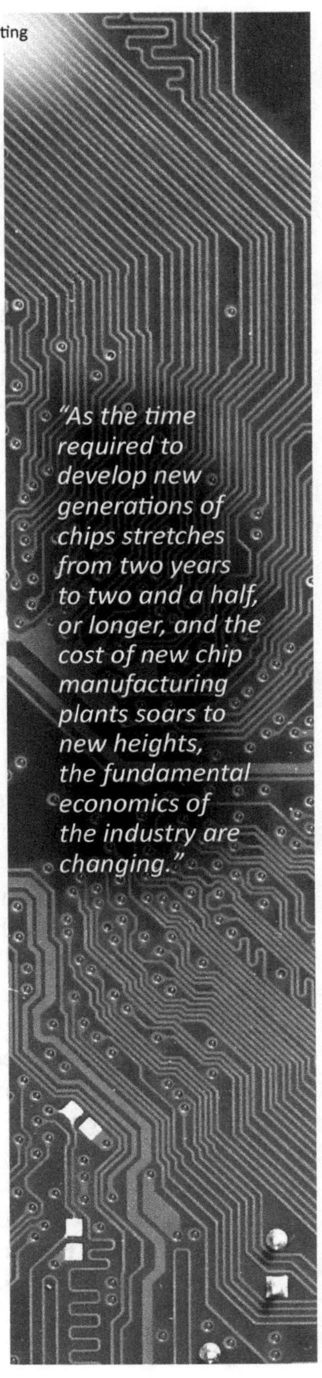

configurations also create complex schedule risks and logistics challenges for chipmakers." Further complicating the economics for would-be chipmakers is the fact that M&A activity went up significantly in 2015, "fueled in part by historically cheap financing that enabled more than $4 trillion of worldwide corporate deal making. Semiconductor M&A also reached unprecedented levels last year, with chip-sector deals worth a combined $117.1 billion announced. This figure is more than five times the $19.9 billion total value of transactions in 2014."

As Nadel explains, "we find that investments in software continue to grow, but in microhardware, this is not the case. The margin erosion is severe, investments in the costs and design and fabrication are now so enormous that the notion of 'build it once and reap the benefit' is one that is not sustainable as costs go up and margins go so far down." To counter this trend, he says, the only real options lie in custom ASIC and FPGAs, but more broadly in the future of open source hardware.

The traditional business for FPGAs in storage and networking will continue to get a boost from new approaches, including NVME drives over Ethernet fabric and innovations in terms of the storage memory hierarchy. Gone are the days when simple DDR-based DIMMS are main memory pushed with hard drives for storage, so with SSD and non-volatile memory computing closer to the compute side, there are more segments of the network, storage, and compute side that can work in harmony with FPGA acceleration.

FPGAs have been important elements of many networking and storage systems where the volumes are not high enough to justify the creation and etching of a full-blow ASICs, and there has been decades of experimentation and niche use of FPGAs as either compute engines or as coprocessors for general purpose CPUs, particularly in the financial services and oil and gas industries. The obvious question is why is Intel, which has fought so hard to bring a certain level of homogeneity in the datacenter, now willing to spend so much money to acquire a company that makes FPGAs as well as hybrid ARM-FPGA devices? What is it that Altera has that Intel thinks it needs, and needs so badly, to make the largest acquisition in its history?

The short answer is a lot of things, and with what Altera brings to the table Intel is hedging its bets on the future of computing in the datacenter – meaning literally serving, storage, and switching – and perhaps in all manner of client devices, too.

The first hedge that Intel is making with the Altera acquisition is that a certain portion of the compute environment that it more or less owns in the datacenter will shift from CPUs to FPGAs.

Intel plans to create a hybrid Atom-FPGA product aimed at the so-called Internet of Things telemetry market, and this will be a monolithic design as well, according to Krzanich; the company is right now examining whether it needs an interim Atom and FPGA product that shares a single package but are not etched on a single die.

Neither Krzanich nor Intel CFO Stacy Smith talked about how much of the CPU workload in the datacenter space they expected to see shift from CPUs to FPGAs over the coming years, but clearly the maturity of the software stack for programming FPGAs is now sufficient that Intel believes such a shift will inevitably happen. Rather than watch those compute cycles go to some other company, Intel is positioning itself to capture that revenue. Presumably the revenue opportunity is much larger than the current revenue stream from Altera, which had $1.9 billion in sales last year. (Intel's own Data Center Group had $14.4 billion in sales, by comparison.)

The chart we did not see as part of the announcement of the Altera deal is what Intel's datacenter business looks like if it does not buy Altera. Intel, as you might expect, wants to focus on how Altera expands its addressable market, which it no doubt does. But there seems little doubt that Intel expected for FPGAs to take a big bite out of its Xeon processor line, too, even if the company doesn't want to talk about that explicitly.

Krzanich said that the future monolithic Xeon-Stratix part would offer more than 2X the performance of the on-package hybrid CPU-FPGA parts it would start shipping in 2016, but what he did not say is that customers using FPGAs today can see 10X to 100X increases in the acceleration of their workloads compared to running them on CPUs. Every one of those FPGAs represents a lot of Xeons that will not get sold, just as is the case in hybrid CPU-GPU machines. (In the largest parallel hybrid supercomputers, there is generally one CPU per GPU, so you might be thinking you only lose half, but the GPUs account for 90 percent of the raw compute capacity, so really a CPU-only machine of equivalent performance would have a factor of 10X more CPUs.) And exacerbating this compression of compute capacity is the fact that Intel will tie the FPGA to the Xeon and makes the combination even faster and with better performance per watt.

The other hedge is that Intel will be able to bring a certain level of customization to its Xeon and Atom product lines without having to provide customizations in the Xeon and Atom chips as it has been doing for the past several years to keep cloud builders, hyperscalers, and integrated system manufacturers happy. Rather than doing custom variants of the chips – in some cases, Intel has as many custom SKUs of a Xeon chip as it has in the standard product line – for each customer, Intel will no doubt suggest that custom instructions and the code that uses them can run in the FPGA half of a Xeon-Stratix hybrid. In some cases, those getting custom parts from Intel are designing chunks of the chip themselves, or they have third parties doing it, according to Krzanich."This does give them a playground," Krzanich said.

The final hedge that Intel is making is that it can learn a bit about making system-on-chip designs and can, if necessary, quickly ramp up an ARM processor business if and when 64-bit ARM chips from Broadcom (soon to be part of Avago Technology), Cavium Networks, Qualcomm, AMD, and Applied Micro take off in the datacenter. In its press release announcing the Altera deal, Intel said that it would operate Altera as a separate business unit and that it would continue to support and develop Altera's ARM-based system-on-chip products; on the call, Krzanich said that Intel was not interested in moving Altera's existing Stratix FPGAs and Arria ARM-FPGA hybrids to its fabs and was happy to leave them at Taiwan Semiconductor Manufacturing Corp.

We will be doing some more analysis of the implications of this Altera deal for future systems and their applications. Stay tuned. And don't be surprised if Avago, which is in the process of acquiring Broadcom for a stunning $37 billion, starts looking real hard at FPGA maker Xilinx.

For device manufacturers, most notably Intel, the integration story gets interesting—and is lighting a fire among other FPGA makers, including Altera rival, Xilinx.

From the relatively recent Intel acquisition of Altera by chip giant Intel, to less talked-about advancements on the programming front (OpenCL progress, advancements in both hardware and software from FPGA competitor to Intel/Altera, Xilinx) and of course, consistent competition for the compute acceleration market from GPUs, which dominate the coprocessor market for now.

At the 2016 Open Compute Summit we finally got a glimpse of one of the many ways FPGAs might fit into the hyperscale ecosystem with an announcement that Intel will be working on future OCP designs featuring an integrated FPGA and Xeon chip. Unlike what many expected, the CPU mate will not be a Xeon D, but rather a proper Broadwell EP. As seen below, this appears to be a 15-core part (Intel did not confirm, but their diagram makes counting rather easy) matched with the Altera Arria 10 GX FPGAs.

This is not a first look at what many expect in the future, which is an FPGA and CPU on a single die—these are in the same package (two chips side by side on a single socket), which begs the question of how the Xeon and FPGA are connected, although one might make a reasonable guess at the same Quick Path Interconnect (QPI) links that are used to link multiple CPUs together so they can share memory and work. Intel is not prepared to comment on that yet, but they are bolstering one of the most important pieces of this story together now, which is on the software and programmability front.

According to Intel's lead for accelerated computing, Eoin McConnell, this configuration created the best balance between CPU and FPGA performance but what is really needed now are the requisite libraries and programming tools to begin to build out a richer ecosystem. Although the hardware story is compelling, the more important bit here is that Intel is boosting its own library prowess for the future of its FPGA hybrids. For instance, as Jason Waxman described at the Open Compute Summit, there is a set of RTL libraries they are working on now and sending to the community for input. The goal is to create this library collection so that users can, in theory anyway, take their FPGA and suddenly have an SSL encryption accelerator or a machine learning library accelerator—all on the fly and with the ability to tweak and tune.

This capability is nothing new, of course. Some companies, including Convey Computing a couple of years ago took the same concept and gave their systems "personalities" that were tunable through libraries and sold as a system versus libraries. It is possible that Intel may, in the future, package such libraries up for distribution and have a business around this to complement their FPGA-CPU chips, although of course, like everything in FPGA land, that remains to be seen.

As for the FPGA libraries Intel is developing into a suite for their forthcoming FPGA push, they "will help users accelerate their workloads and give developers a platform to start with. With the these libraries, we've looked at a range of different acceleration demands and end user demands to put together a suite. We can't say what all is in it yet, but expect a range of standard acceleration for a number of segments, including cloud, networking and traditional enterprise," McConnell says. "The goal is to provide the right suites to help people use the FPGAs now for things like compression, encryption, and visualization and we're continuing to work with Altera and what we now call the Programmable Solutions Group to look at other use cases."

As we lead into the 2017 timeframe when the Broadwell EP/FPGA hybrids become available, there will be other work going on to bolster the programmability and tooling for similar devices, including (our guess) a Xeon D-matched part. McConnell says Intel is getting a great deal of interest in what they might be able to do for a range of applications where we see FPGAs already—and some new areas, including machine learning.

There will continue to be a story for an Intel as well as ARM for system vendors, but we should also mention another very important company in the future of FPGAs—and that is IBM. The company has put a great deal of its acceleration focus on GPUs and the right interconnects to offer high performance, but reconfigurable computing is not off Big Blue's radar.

IBM did not just stake the future of its Power chip and the systems business on which it depends on the OpenPower Foundation, a consortium now with 160 members after more than two years of cultivation by Big Blue and its key early partners – Google, Nvidia, Mellanox Technologies, and Tyan. It has staked its future in the systems business on the idea of accelerated computing, which means using a mix of processors and accelerators to maximum performance and minimize costs and thermals for specific workloads.

It is hard to argue that the future of computing in the datacenter is moving away from general purpose processors and systems (perhaps with the exception of hyperscalers, who need homogeneity to keep the acquisition and operation costs low so they can grow their customer bases and services) to more specialized gear. Look at how many different kinds and types of processors that Intel sells just to see the diversity, which is about to be expanded with Altera FPGAs once Intel completes the $16.7 billion acquisition of that chip maker.

In the OpenPower camp, IBM and Nvidia have been tag teaming with hybrid computing for a while, with the big wins being the future "Summit" and "Sierra" supercomputers that the U.S. Department of Energy is spending $325 million to build for Oak Ridge National Laboratory and Lawrence Livermore National Lab, respectively. These are pre-exascale systems based on the future Power9 chips from IBM lashed to the future "Volta" Tesla GPU coprocessors from Nvidia, with the CPUs and GPUs linked by NVLink high-speed interconnects and the resulting hybrid nodes linked by EDR InfiniBand (or perhaps HDR if it gets completed in time).

Naturally and predictably, in the wake of the Intel-Altera deal, Xilinx is becoming an important member of the OpenPower compute platform and at the has announced a strategic collaboration with IBM under the auspices of the OpenPower Foundation to more tightly couple its FPGAs with Power processors and ultimately Nvidia GPUs where that is appropriate. This multi-faceted hybrid computing is something that we expected, and we said as much back in March 2016 when we attended the first OpenPower Summit in Silicon Valley, and in fact, we used the high frequency trading applications from Algo-Logic, which bring all three technologies together, as an example of how this could work.

John Lockwood, CEO at Algo-Logic, summed it up this way: FPGAs are deployed where you need low latency on transactions, GPUs are used where you need high throughput calculations for the parts of the application components that can be parallelized, and CPUs are used for those portions of the code need fast execution on single threads. The trick is making it all work together to accelerate *the entire application*. This is not as simple as buying a custom processor as any type, but the OpenPower partners think the arguments are compelling for this approach. Compelling enough for IBM to sell off its System x business and essentially stake its system future on the idea.

The question we had when being briefed about the OpenPower announcements at the Supercomputing Conference was this: When does accelerated computing become normal? And then we thought about a future where

"With every generation – and it is becoming more and more interesting math – we move functions and accelerators onto the chips. With each of our chips, we go through a longer and longer list of special purpose accelerators. When is it going to be that the math will be right to pull in a general purpose programmable accelerator onto the chip? Is it at 7 nanometers or 3 nanometers when that is going to happen? I don't know just yet."

all of this technology – CPU, GPU, and FPGA – might end up in a single package or on a single die anyway as Moore's Law progresses for the next decade or so.

Brad McCredie, who is vice president Power Systems development at IBM and president of the OpenPower Foundation, offers his ideas on these issues and when everything that can be accelerated is accelerated.

"I think what that comes down to is predicting a rate of change in software," McCredie explains. "That is going to be the gate to the timeline that you are thinking about. We see that these transitions do take five to ten years for software to migrate, but I do think that is an end state and people may choose to debate that with me. I think we will see that accelerated computing will be the norm and that software will be developed that way. Of course we are going to build lots of tools and aids to make it easier and easier to use these hybrid architectures. But this is going to be the new normal and we are going there."

With Intel already having integrated GPUs on selected Xeon processors (both on-package and on-die variations), and talking about how it will have in-package and eventually on-die FPGA accelerators on selected members of the Xeon family, it is natural enough to ask if the OpenPower partners will ever work together to create various integrated CPU-FPGA or CPU-GPU or even CPU-GPU-FPGA hybrid chips. There may be some technical barriers to this that need to be hurdled, of course, but it is not a ridiculous thing to contemplate – particularly as Moore's Law starts running out of gas. Here is what McCredie had to say about that:

"With every generation – and it is becoming more and more interesting math – we move functions and accelerators onto the chips. With each of our chips, we go through a longer and longer list of special purpose accelerators. When is it going to be that the math will be right to pull in a general purpose programmable accelerator onto the chip? Is it at 7 nanometers or 3 nanometers when that is going to happen? I don't know just yet. The one thing that I would point out is that as we are getting more and more out of the accelerators, the truth is the amount of silicon in a system that is being devoted to accelerators is outstripping CPUs in many cases. So maybe these things are going to stay separate for quite a while, only because we are just going to add more and more silicon into the system to get the job done as Moore's Law slows down. When that happens, the corollary is that you need more square millimeters of silicon to get the job done."

As we have pointed out a number of times in 2016 (and before) at *The Next Platform,* one could make an argument for a central processor complex comprised of CPU cores with fast single-thread performance that have had their vector math units ripped out that is coupled to an on-package GPU and FPGA. But McCredie pushed back on that idea.

"The idea is absolutely not stupid, and these are trajectories that we could go on," says McCredie. "But for several generations, right now where we are sitting in the industry, as far as scaling and performance goes, the investment is key is in the bus and the communication between the processor and the accelerators. This is where the differentiation is going to take place. We need a lot of silicon to do the processor and the accelerators, and no one is saying that they need less CPU or less acceleration. We see demand for more of both, and so need to get better and much more efficient communication between these components."

For OpenPower, that means a few different things. First, it means embracing and enhancing the Coherent Accelerator Processor Interface (CAPI) that is part of the Power8 chip and that allows for coherent memory access between the Power8 processor and accelerators that link over the PCI-Express bus in the system. As part of the multi-year agreement between IBM and Xilinx, the two will be working on CAPI integration for Xilinx FPGAs, the SDAccel programming stack for FPGAs will be ported to Power processors and optimized for the combination, and Xilinx roadmaps will be aligned with the combined roadmap from the OpenPower partners so compute and networking in their various forms in these hybrid machines move together in unison and at a predictable pace. The Xilinx-IBM agreement also includes joint marketing and sales efforts, too.

The important thing to note is that the Power8+ chip due in 2016 will have both NVLink ports for boosting the bandwidth and lowering the latency between Power chips and "Pascal" GP100 Tesla coprocessors as well as the existing CAPI links for talking to other kinds of accelerators such as FPGAs.

In 2017, IBM will move to the Power9 processors and both CAPI and NVLink will be enhanced to create the foundational technology for the Summit and Sierra systems for the Department of Energy; the enhanced NVLink will be used to hook the Power9 chips to the "Volta" GV100 Tesla coprocessors. The roadmap calls for HDR InfiniBand and matching adapters running at 200 Gb/sec linking the hybrid nodes to each other. The precise generations for Xilinx chips is not clear yet – they just inked the deal, after all. At the moment, IBM and FPGA partners (including Altera and Xilinx) have been able to create a CAPI-enabled PCI-Express port on the FPGA out of logic gates on the FPGA itself, which is something you cannot do with other chips because they are not malleable like FPGAs. In a future generation, the CAPI bus will be made more robust, says McCredie, and will be "pulled away from being 100 percent tied to PCI-Express," as he put it. Xilinx will align with this enhanced CAPI bus, but don't expect for Nvidia Tesla GPU coprocessors to use it. The rule is NVLink for GPUs, CAPI for everything else. (We suspect that Enhanced NVLink will have a maximum of eight ports per device instead of four and run at a higher clock speed that 20 GB/sec that the original NVLink has. Oh, and the Power9 chip will have a new microarchitecture and use a new chip process (14 nanometers) at the same time, by the way.

This is a lot of change, but that is precisely what the OpenPower partners have signed up for to chase exascale computing.

IBM is also talking about its own use of Tesla K80 accelerators on two-socket "Firestone" systems that underpin its Watson cognitive computing stack. McCredie tells *The Next Platform* that IBM is accelerating various deep learning algorithms in the Watson stack using the Tesla GPUs, and as proof points IBM says that the Retrieve and Rank APIs in the Watson stack have been accelerated by a factor of 1.7X and on natural language processing jobs the performance has been goosed by a factor of 10X. McCredie said that IBM had not yet deployed either the Tesla M4 or Tesla M40 GPU accelerators, announced last week by Nvidia, underneath Watson, but that given their aim at machine learning, he expected that IBM would give them consideration.

To help more customers make the move, the SuperVessel Power cloud that IBM set up in China earlier in 2016 has expanded GPU and FPGA acceleration, and IBM's own centers in Poughkeepsie, New York and Montpelier, France have been beefed up, too. And of course, Oak Ridge and Lawrence Livermore will be doing development work on hybrid Power-GPU setups, too.

The wins for Power-Tesla hybrid computing and the test beds for Power-FPGA computing have been documented in *The Next Platform*, and what we wanted to know is how the uptake is going outside of these HPC labs and the oil and gas industry where this idea was originally rejected and then took off. Rice University, Baylor University, Oregon State, and the University of Texas all have hybrid clusters for doing research, and Louisiana State University is doing on with FPGA accelerated Power clusters. McCredie says that there are examples of companies doing network function virtualization and running other workloads underway at telecommunication firms and service providers, and that eight big proofs of concept are underway in various large enterprise accounts.

The reason is simple: Like Google, they have to beat Moore's Law, any way that they can. That is why Google was a founding member of the OpenPower Foundation, after all.

One aspect of the market that we will continue to watch is how smaller vendors tackle the market with FPGA-based systems that can abstract away some of the complexity and get full performance out of the newest reconfigurable devices.

What is interesting, however, is that as this "old" 1980s technology has resurfaced over the last few years in the wake of new data-intensive problems, the performance has been bolstered, but the programmability problem persists, even with the work that has been done to tie FPGAs to more common approaches like OpenCL. But despite the barriers, FPGAs are exploiting new opportunities in data analytics, giving a second wind to an industry that seemed to languish on the financial services, military and oil and gas fringes.

Backed by Altera, Xilinx, Nallatech, and other vendors and with hooks into OpenPower, the gates have opened for FPGAs to prove their mettle in a host of new web-scale datacenter environments, including Microsoft for its Bing search service. And it's this type of hyper-dense search, along with neural networks, deep learning, and other massive machine learning applications, where they stand to shine. But again, back to reality, while the core value of FPGAs is that they are reconfigurable on the fly, this is also an Achilles heel, programmatically speaking. No matter how robust, high performance, and low power, if they cannot be programmed without specialized staff, what's the point? Now, with Intel/Altera, and momentum in the OpenPower sphere to bring FPGAs to a wider market, the impetus is greater than ever to make FPGAs available programmatically. But there could be a way of wrapping around that problem, at least for some algorithms.

Before we get to that, the Bing example along with the other financial and energy exploration codes where FPGAs have historically been found all share something else in common. The FPGAs are tied to a traditional sequential X86 processing environment that is based on clustering for scale and added performance. Further, the addition of FPGAs, while adding performance for specific problems, adds additional programming complexity, With this in mind, it's not difficult to imagine building a supercomputer from these FPGAs, but according to Patrick McGarry, VP of engineering at FPGA-based data analytics server maker, Ryft, one of the main goals is to move away from both an X86 and clustering mindset and make the FPGA the analytics workhorse. And what's notable here is that they may have made the programmability leap.

"We set out to design a 1U box that could minimize or even

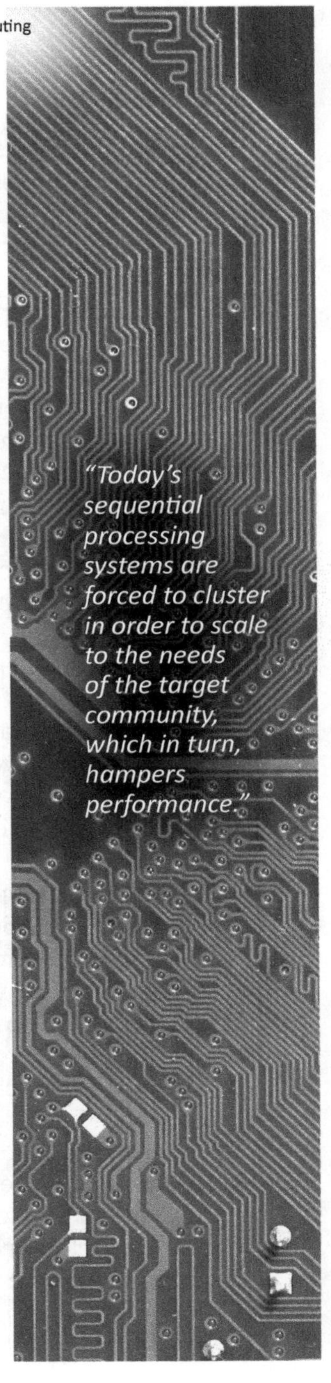

eliminate the need for clustering for the majority of an enterprise's analytics problems," explained McGarry. The Ryft ONE boxes his company makes look like a network appliance from the outside, but are outfitted with a X86 processor running Linux with no special software mojo other than the drivers for the FPGA. As a side note, the company ran a series of internal benchmarks on both Altera and Xilinx FPGAs and found that Xilinx was the clear winner for their specific needs based on how quickly it could be reconfigured. McGarry said they tested a few others, but their work with Xilinx has allowed to do something very interesting indeed—to develop a custom interconnect, which is their secret sauce, that allows for some pretty impressive results (at least based on their own benchmarks). "Xilinx does not know what we are putting in their FPGAs. It's the interconnect, the logic, and how we're passing data. There is no third-party interface or development platform that can do what we did, they've been a great partner but this is all a unique approach."

So what we have here is an FPGA-based system where the host processor handles basic tasks, allowing the FPGA to exclusively handle the bulk of the processing–all the while minimizing data movement, even when compared to how the fastest in-memory analytics approaches do so (Spark for instance). The numbers below are done within a 700 watt power envelope for the entire box. The efficiencies come in part from the FPGA, but this device is also the way the Ryft ONE it cuts down on data movement.

"Current large-cluster solutions using tools like Hadoop and Spark require an inordinate amount of ETL (extract, transform, and load) work. This often artificially inflates the size of the data significantly. And making matters worse, users are often forced to index the data to be able to effectively search it, which can further artificially inflate the data set size. (Not to mention the amount of time it takes to do those things – days and weeks in some cases, especially if multiple indexes are required.) With the Ryft ONE, you can analyze data in its rawest form," McGarry highlighted. "You aren't required to ETL or index it. That's a huge differentiator. If you want to add some structure for your own purposes, you are certainly more than welcome (such as perhaps XML, or something along those lines), but it is by no means a requirement."

If your core business is machine learning, certain types of search functions, or in coming months, image and video processing, this might be of interest. And while you can still make use of that unnamed but Intel "high performance" X86 processor inside and offload things to the FPGA on board, you could, but it would be a damned expensive way to do things (these are sold as hosted/on-prem rentals with an $80,000 configuration fee and $120,000 per year license that includes all updates, new primitives, etc.).

But of course, back the big question here—the efficiencies can be benchmarked in theory, but how do you actually program the thing, especially since this is one of the limiting factors for broader FPGA adoption? And further, how is it this small company claims to have figured out the secret to making FPGAs simple to program and use when the largest makers and their communities haven't solved this problem?

The answer brings us back to the fact that this is, at its simplest, an appliance. It comes with "precooked" ways of handling a select set of highly valuable data analytics problems. But as McGarry added, the API Ryft is using is open (available on its site) and based on C, which means in theory, it's possible to wrap almost any programming language in the world around it since many (Java, Scala, Python, etc) can all invoke C. The idea then is that the only thing required to use the FPGA is to make a high level function call—you call the routine and it's there.

"We tried to abstract away everything possible and by doing it this way, you don't even need to know how to spell FPGA, and it also makes it easier on our end because we don't have to support all of the other tools (visualization, etc), you can just write it. Our box is for performance purely."

So with these potential advantages in mind in terms of programmability and performance, how can such systems—which again, were not built with traditional clustering and scalability in mind—still scale? The target workloads require processing of around 10TB on average, which is one of the reasons they decided to natively support up to 48TB in the 1U Ryft ONE. While it's possible to do that across multiple systems, overstocking the machine means most of their users will never need to, in theory at least.

And even with all the data processing capability in the world, it's meaningless without a way to move the data. The Ryft boxes have two 10 Gb/sec Ethernet ports that can be used for arbitrary data ingress and egress. "Since our analytics performance in our backend analytics fabric is on the order of 10 gigabytes per second, we can easily handle line rate 10 gigabit Ethernet (since using 8B10B encoding, that translates to only one gigabyte per second) – with room to spare. This means that should the market decide that we need higher network interface speeds, we can make that happen with a minimal amount of re-engineering."

In terms of scaling beyond this, McGarry says that if a user needs more than the 48TB for data at rest, and more than two 10 Gb/sec network links, then you could utilize multiple Ryft boxes to scale. "However," he notes, "you wouldn't use them in what you traditionally think of as a 'cluster'. You would instead implement a data sharding approach, determining which of the two (or more) Ryft ONE boxes you'd send the data for processing."

Other companies dipped an early toe in the FPGA

box waters, including high performance computing server maker, Convey, which did something similar for a select set of workloads. As OpenPower continues its development, we can expect a new crop of FPGA-powered systems to emerge. Further, if indeed it's true that Intel will be buying Altera, one can only guess what lies on the horizon, but for now, we'll wait in the wings for a user story to emerge from Ryft to provide some much-needed real-world insight.

Other companies that see the writing on the wall for the future of FPGAs include those that monitor the performance differences between existing and future general purpose processors and accelerators and try to tack FPGAs on for additional boost without the complexity.

Nallatech doesn't make FPGAs, but it does have several decades of experience turning FPGAs into devices and systems that companies can deploy to solve real-world computing problems without having to do the systems integration work themselves.

With the formerly independent Altera, now part of Intel, shipping its Arria 10 FPGAs, Nallatech has engineered a new coprocessor card that will allow FPGAs to keep pace with current and future Tesla GPU accelerators from Nvidia and "Knights Landing" Xeon Phi processors and coprocessors from Intel. The architectures of the devices share some similarities, and that is no accident because all HPC applications are looking to increase memory bandwidth and find the right mix of compute, memory capacity, and memory bandwidth to provide efficient performance on parallel applications.

Like the Knights Landing Xeon Phi, the new 510T uses a mix of standard DDR4 and Hybrid Memory Cube (HMC) memory to provide a mix of high bandwidth, low capacity memory with high capacity, relatively low bandwidth memory to give an overall performance profile that is better than a mix of FPGAs and plain DDR4 together on the same card.

In the case of the 510T card from Nallatech, the compute element is a pair of Altera Arria 10 GX 1150 FPGAs, which are etched in 20 nanometer processes from foundry partner Taiwan Semiconductor Manufacturing Corp. The higher-end Stratix 10 FPGAs are made using Intel's 14 nanometer processes and pack a lot more punch with up to 10 teraflops per device, but they are not available yet. Nallatech is creating coprocessors that will use these future FPGAs. But for a lot of workloads, as Nallatech president and founder Allan Cantle explains to *The Next Platform*, the compute is not as much of an issue as memory bandwidth to feed that compute. Every workload is different, so that is no disrespect to the Stratix 10 devices, but rather a reflection of the key oil and gas customers that Nallatech engaged with to create the 510T card.

"In reality, these seismic migration algorithms need huge amounts of compute, but fundamentally, they are streaming algorithms and they are memory bound," says Cantle. "When we looked at this for one of our customers, who was using Tesla K80 GPU accelerators, somewhere between 5 percent and 10 percent of the available floating point performance was actually being used and 100 percent of the memory bandwidth was consumed. That Tesla K80 with dual GPUs has 24 GB of memory and 480 GB/sec of aggregate memory bandwidth across those GPUs, and it has around 8.7 teraflops of peak single precision floating point capability. We have two Arria 10s, which are rated at 1.5 teraflops each, which is just around 3 teraflops total but I think the practical upper limit is 2 teraflops., but that is just my personal take. But when you look at it, you only need 400 gigaflops to 800 gigaflops, making very efficient use of the FPGA's available flops, which you cannot do on a GPU."

The issue, says Cantle, is that the way the GPU implements the streaming algorithm at the heart of the seismic migration application that is used to find oil buried underground, it makes many accesses to the GDDR5 memory in the GPU card, which is what is burning up all of the memory bandwidth. "The GPU consumes its memory bandwidth quite quickly because you have to come off chip the way the math is done," Cantle continues. "The opportunity with the FPGA is to make this into a very deep pipeline and to minimize the amount of time you go into global memory."

The trick that Nallatech is using is putting a block of HMC memory between the two FPGAs on the board, which is a fast, shared memory space that the two FPGAs can actually share and address at the same time. The 510T is one of the first compute devices (rather than networking or storage devices) that is implementing HMC memory, which has been co-developed by Micron Technology and Intel, and it is using the second generation of HMC to be precise. (Nallatech did explore first generation HMC memory on FPGA accelerators for unspecified government customers, but this was not commercially available as the 510T card is.)

In addition to memory bandwidth bottlenecks, seismic applications used in the oil and gas industry also have memory capacity issues. The larger the memory that the compute has access to, the larger the volume (higher number of frequencies) that the seismic simulation can run. With the memory limit on a single GPU, says Cantle, this particular customer was limited to approximately 800 volumes (it is actually a cube). Oil and gas customers would love to be able to do 4K volumes (again cubed), but that would require about 2 TB of memory to do.

So the 510T card has four ports of DDR4 main memory to supply capacity to store more data to do the larger and more complex seismic analysis, and by ganging up 16 cards together across hybrid CPU-FPGA nodes, Nallatech can break through that 4K volumes barrier and reach the level of performance that oil and gas companies are looking for.

The HMC memory comes in 2 GB capacity, with 4 GB being optional, and has separate read and write ports, each of which deliver 30 GB/sec of peak bandwidth per FPGA on the card. The four ports of DDR4 memory that link to the other side of the FPGAs deliver 32 GB of capacity per FPGA (with an option of 64 GB per FPGA) and 85 GB/sec of peak bandwidth. So each card has 290 GB/sec of aggregate bandwidth and 132 GB of memory for the applications to play in.

These FPGA cards slide into a PCI-Express x16 slot, and in fact, Nallatech has worked with server maker Dell to put these into a custom, high-end server that can put four of these cards and two Xeon E5 processors into a single 1U rack-mounted server. The Nallatech 510T cards cost $13,000 each at list price, and the cost of a server with four of these plus an OpenCL software development kit and the Altera Quartus Prime Pro FPGA design software added to is $60,000.

Speaking very generally, the two-FPGA card can deliver about 1.3X the performance of the Tesla K80 running the seismic codes at this oil and gas customer in about half the power envelope, says Cantle and there is a potential upside of 10X performance for customers that have larger volume datasets or who are prepared to optimize their algorithms to leverage the strengths of the FPGA. But Nallatech also knows that FPGAs are more difficult to program than GPUs at this point, and is being practical about the competitive positioning.

"At the end of the day, everyone needs to be a bit realistic here," says Cantle. "In terms of price/performance, FPGA cards do not sell in the volumes of GPU cards, so we hit a price/performance limit for these types of algorithms. The idea here is to prove that today's FPGAs are competent at what GPUs are great at. For oil and gas customers, it makes sense for companies to weigh this up. Is it a slam dunk? I can't say that. But if you are doing bit manipulation problems – compression, encryption, bioinformatics – it is a no brainer that the FPGA is far better – tens of times faster – than the GPU. There will be places where the FPGA will be a slam dunk, and with Intel's purchase of Altera, their future is certainly bright."

The thing we observe is that companies will have to not look just at raw compute but how their models can scale across the various memory in a compute element and across multiple elements lashed together inside of a node and across nodes.

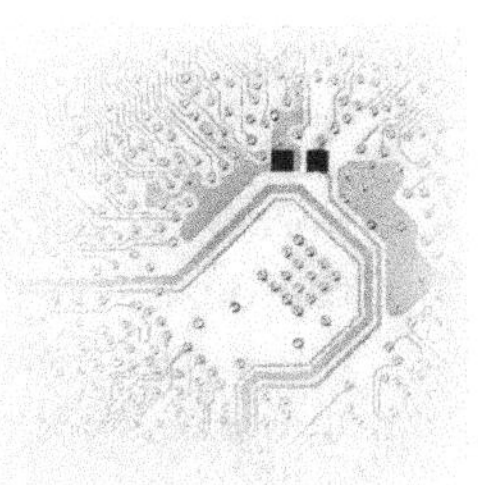

CONCLUSION

It would be far easier on the IT industry if Denard scaling and Moore's Law improvements in the economics of chips had just continued apace through the end of the 2000s and into today. General purpose CPUs – and that generally means the X86 instruction set these days – would have reigned supreme and other architectures of processors and coprocessor would have been driven out of the market by volume economics and the homogeneity of the application and system software stacks.

But improvements in chip manufacturing processes are slowing, and that means that many things they were old, like FPGAs, are new again. To get the most efficiency, HPC centers, hyperscalers, and cloud builders have to look at every possible option to accelerate the performance of their applications – even those that are somewhat difficult to program, as the FPGA can be, even with modern tools to help bridge the gap between traditional programming languages such as C or C++ and the Verilog/VHDL language that describes the logic implemented on FPGAs that turns them from a pile of logic gates into a kind of hardware-software hybrid that has the benefits of both.

If there is a lesson from the modern era, where we are seeing a kind of Cambrian Explosion in the variety of types of compute elements in the datacenter, it is that compute monoculture, which has provided so many good benefits to end user companies (as well as X86 chip supplier, Intel), is not sustainable. A mix of compute types within a system as well as across systems running various parts of the application workflow is necessary to achieve the maximum efficiency and build the best possible system. The history of the FPGA and its relatively modest adoption in the enterprise is not necessarily indicative of the future adoption of FPGAs within future systems. We believe that FPGAs will find their place, just as GPUs have, alongside CPUs in hybrid systems, and for the same reason that network devices have had a mix of ASICs and FPGAs for a long time (and many even have general-purpose CPUs to run network applications, too).

There are many reasons why we believe FPGAs will see broader and deeper adoption, but perhaps one of the most important reasons is that they are malleable hardware, a kind of device that is somewhere between an ASIC that does one and only one thing and a CPU that can be programmed to do anything but which has a fairly deep software stack that needs to be kept up. Ironically, this malleability is something that FPGA makers like Xilinx and Altera tried to downplay in the early days of the commercialization of these devices. FPGAs, thanks to their relatively low volumes, are also ideal testbeds for new chip making technologies.

As making smaller and smaller circuits to drive capacity improvements on CPUs gets increasingly difficult because we are approaching the limits of physics with CMOS technology, there will perhaps be more importance placed on creating systems that are based at least in part on FPGAs and that will themselves be more malleable but also be more precise than a CPU, which needs software to tell it how to behave, becomes harder and harder.

While it would be interesting to live in a world where FPGAs killed off CPUs and GPUs for compute, it seems unlikely that the Moore's Law pressures are so great to push most applications off CPUs or GPUs and onto FPGAs. But some applications will make that jump, and in other cases, the offload model and the tools supporting it will allow for some portions of an application to be moved to FPGAs, and we think that is precisely what many companies will do.

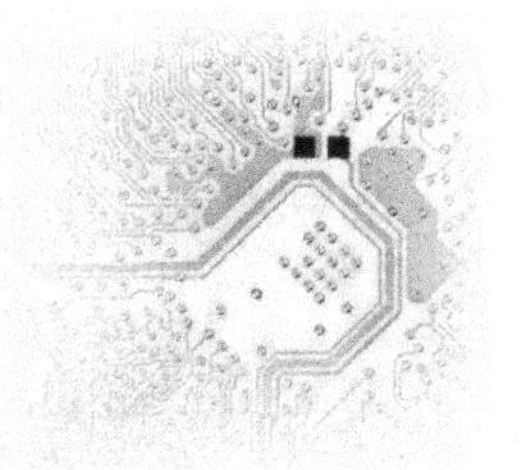

References

Note that much of the material in this book has appeared in The Next Platform from late 2015 until the beginning of 2017. Searching by keyword on the site or exact phrase will bring desired article with all links included in web format.

Abdelfattah, Mohamed S., Andrei Hagiescu, and Deshanand Singh. "Gzip on a chip: High performance lossless data compression on fpgas using opencl." *Proceedings of the International Workshop on OpenCL 2013 & 2014*. ACM, 2014.

Andrews, David, and Marco Platzner. "Programming models for reconfigurable manycore systems." *Reconfigurable Communication-centric Systems-on-Chip (ReCoSoC), 2016 11th International Symposium on*. IEEE, 2016.

Brown, Stephen, and Jonathan Rose. "FPGA and CPLD architectures: A tutorial." *IEEE design & test of computers* 13.2 (1996): 42-57.

Chiou, Derek. "Intel Acquires Altera: How Will the World of FPGAs be Affected?." *Proceedings of the 2016 ACM/SIGDA International Symposium on Field-Programmable Gate Arrays*. ACM, 2016.

Choi, Young-kyu, et al. "A quantitative analysis on microarchitectures of modern CPU-FPGA platforms." *Proceedings of the 53rd Annual Design Automation Conference*. ACM, 2016.

Chow, Paul. "An Open Ecosystem for Software Programmers to Compute on FPGAs." *FSP 2016; Third International Workshop on FPGAs for Software Programmers; Proceedings of*. VDE, 2016.

Crespo, Maria Liz, et al. "Reconfigurable Virtual Instrumentation Based on FPGA for Science and High-Education." *Field-Programmable Gate Array (FPGA) Technologies for High Performance Instrumentation*. IGI Global, 2016. 99-123.

D'Hollander, Erik. "High-level synthesis for FPGAs, the Swiss army knife for high-performance computing." *Programming models for FPGA*. Maison de la Simulation, Paris, France, 2016.

Dang, Huynh Tu, et al. "Network Hardware-Accelerated Consensus." *arXiv preprint arXiv:1605.05619* (2016).

De La Piedra, Antonio, An Braeken, and Abdellah Touhafi. "Sensor systems based on FPGAs and their applications: A survey." *Sensors* 12.9 (2012): 12235-12264.

Denisenko, Dmitry. "OpenCL Compiler Tools for FPGAs." *Proceedings of the 4th International Workshop on OpenCL*. ACM, 2016.

DiCecco, Roberto, et al. "Caffeinated FPGAs: FPGA Framework For Convolutional Neural Networks." *arXiv preprint arXiv:1609.09671* (2016).

Escobar, Fernando A., Xin Chang, and Carlos Valderrama. "Suitability Analysis of FPGAs for Heterogeneous Platforms in HPC." *IEEE Transactions on Parallel and Distributed Systems* 27.2 (2016): 600-612.

Feng, Liang, et al. "HeteroSim: A Heterogeneous CPU-FPGA Simulator." *IEEE Computer Architecture Letters* (2016).

Giefers, Heiner, et al. "Analyzing the energy-efficiency of sparse matrix multiplication on heterogeneous systems: A comparative study of GPU, Xeon Phi and FPGA." *Performance Analysis of Systems and Software (ISPASS), 2016 IEEE International Symposium on*. IEEE, 2016.

Gu, Changyi. "Field-Programmable Gate Arrays." *Building Embedded Systems*. Apress, 2016. 191-231.

How, Dana L., and Sean Atsatt. "Sectors: Divide & Conquer and Softwarization in the Design and Validation of the Stratix® 10 FPGA." *Field-Programmable Custom Computing Machines (FCCM), 2016 IEEE 24th Annual International Symposium on*. IEEE, 2016.

Hannig, Frank. "A Quick Tour of High-Level Synthesis Solutions for FPGAs." *FPGAs for Software Programmers*. Springer International Publishing, 2016. 49-59.

Huang, Muhuan, et al. "Programming and Runtime Support to Blaze FPGA Accelerator Deployment at Datacenter Scale." *Proceedings of the Seventh ACM Symposium on Cloud Computing*. ACM, 2016.

Inggs, Gordon. "Algorithmic Trading: A brief, computational finance case study on data centre FPGAs." *arXiv preprint arXiv:1607.05069* (2016).

Iordache, Anca, et al. "High performance in the cloud with FPGA groups." *Proceedings of the 9th International Conference on Utility and Cloud Computing*. ACM, 2016.

Koch, Dirk, Frank Hannig, and Daniel Ziener. "FPGAs for Software Programmers." (2016).

Koch, Dirk, Daniel Ziener, and Frank Hannig. "FPGA Versus Software Programming: Why, When, and How?." *FPGAs for Software Programmers*. Springer International Publishing, 2016. 1-21.

Kindratenko, Volodymyr. "Novel computing architectures." *Computing in Science & Engineering* 11.3 (2009): 54-57.

Krommydas, Konstantinos, Ruchira Sasanka, and Wu-chun Feng. "Bridging the FPGA programmability-portability Gap via automatic OpenCL code generation and tuning." *Application-specific Systems, Architectures and Processors (ASAP), 2016 IEEE 27th International Conference on*. IEEE, 2016.

Lacey, Griffin, Graham W. Taylor, and Shawki Areibi. "Deep Learning on FPGAs: Past, Present, and Future." *arXiv preprint arXiv:1602.04283* (2016).

Lacey, Griffin James. *Deep Learning on FPGAs*. Diss. 2016.

Lawande, Abhijeet, Alan D. George, and Herman Lam. "An OpenCL framework for distributed apps on a multidimensional network of FPGAs." *Proceedings of the Sixth Workshop on Irregular Applications: Architectures and Algorithms*. IEEE Press, 2016.

Lee, Seyong, Jungwon Kim, and Jeffrey S. Vetter. "Openacc to fpga: A framework for directive-based high-performance reconfigurable computing." *Parallel and Distributed Processing Symposium, 2016 IEEE International*. IEEE, 2016.

Lewis, David, et al. "The Stratix™ 10 Highly Pipelined FPGA Architecture." *Proceedings of the 2016 ACM/SIGDA International Symposium on Field-Programmable Gate Arrays*. ACM, 2016.

Levin, Ilya Izrailevich, et al. "Reconfigurable computer systems: from the first FPGAs towards liquid cooling systems." *Supercomputing frontiers and innovations* 3.1 (2016): 22-40.

Miller, Luke Andrew. "The Role of FPGAs in the Push to Modern and Ubiquitous Arrays." *Proceedings of the IEEE* 104.3 (2016): 576-585.

Monmasson, Eric, et al. "FPGAs in industrial control applications." *IEEE Transactions on Industrial Informatics* 7.2 (2011): 224-243.

Neshatpour, Katayoun, Avesta Sasan, and Houman Homayoun. "Big data analytics on heterogeneous accelerator architectures." *Hardware/Software Codesign and System Synthesis (CODES+ ISSS), 2016 International Conference on*. IEEE, 2016.

Nguyen, Xuan-Thuan, et al. "An efficient FPGA-based database processor for fast database analytics." *Circuits and Systems (ISCAS), 2016 IEEE International Symposium on*. IEEE, 2016.

Nurvitadhi, Eriko, et al. "Accelerating recurrent neural networks in analytics servers: Comparison of FPGA, CPU, GPU, and ASIC." *Field Programmable Logic and Applications (FPL), 2016 26th International Conference on*. IEEE, 2016.

Podobas, Artur, and Mats Brorsson. "Empowering OpenMP with Automatically Generated Hardware." *SAMOS XVI*. 2016.

Rose, Jonathan, Abbas El Gamal, and Alberto Sangiovanni-Vincentelli. "Architecture of field-programmable gate arrays." *Proceedings of the IEEE* 81.7 (1993): 1013-1029.

Sadrozinski, Hartmut F-W., and Jinyuan Wu. *Applications of field-programmable gate arrays in scientific research*. CRC Press, 2016.

Schmid, Moritz, et al. "Big Data and HPC Acceleration with Vivado HLS." *FPGAs for Software Programmers*. Springer International Publishing, 2016. 115-136.

Shen, Hao, and Qinru Qiu. "An FPGA-based distributed computing system with power and thermal management capabilities." *Computer Communications and Networks (ICCCN), 2011 Proceedings of 20th International Conference on*. IEEE, 2011.

Shao, Yakun Sophia. *Design and Modeling of Specialized Architectures*. Diss. Harvard University Cambridge, Massachusetts, 2016.

Sharma, Hardik, et al. "From high-level deep neural models to FPGAs." *Microarchitecture (MICRO), 2016 49th Annual IEEE/ACM International Symposium on*. IEEE, 2016.

Singh, Deshanand, and Peter Yiannacouras. "OpenCL." *FPGAs for Software Programmers*. Springer International Publishing, 2016. 97-114.

Singh, Satnam. "Computing without processors." *Communications of the ACM* 54.8 (2011): 46-54.

Stroobandt, Dirk, et al. "EXTRA: Towards the exploitation of eXascale technology for reconfigurable architectures." *Reconfigurable Communication-centric Systems-on-Chip (ReCoSoC), 2016 11th International Symposium on*. IEEE, 2016.

Syed Waqar, N. A. B. I. "FPGA Port of a Large Scientific Model from Legacy Code: The Emanuel Convection Scheme." *Parallel Computing: On the Road to Exascale* 27 (2016): 469.

Szymanski, Ted H. "Securing the Industrial-Tactile Internet of Things With Deterministic Silicon Photonics Switches." *IEEE Access* 4 (2016): 8236-8249.

Tapiador, R., et al. "Comprehensive Evaluation of OpenCL-based Convolutional Neural Network Accelerators in Xilinx and Altera FPGAs." *arXiv preprint arXiv:1609.09296* (2016).

Varma, B. Sharat Chandra, Kolin Paul, and M. Balakrishnan. *Architecture exploration of FPGA based accelerators for bioinformatics applications*. Vol. 55. Springer, 2016.

Wang, Kui, and Jari Nurmi. "Using OpenCL to rapidly prototype FPGA designs." *Nordic Circuits and Systems Conference (NORCAS), 2016 IEEE*. IEEE, 2016.

Wang, Zeke, et al. "Relational query processing on OpenCL-based FPGAs." *Field Programmable Logic and Applications (FPL), 2016 26th International Conference on*. IEEE, 2016.

Weisz, Gabriel, et al. "A Study of Pointer-Chasing Performance on Shared-Memory Processor-FPGA Systems." *Proceedings of the 2016 ACM/SIGDA International Symposium on Field-Programmable Gate Arrays*. ACM, 2016.

Wu, Chien-Chung, and Kai-Wen Weng. "The Development and Implementation of a Real-Time Depth Image Capturing System Using SoC FPGA." *2016 30th International Conference on Advanced Information Networking and Applications Workshops (WAINA)*. IEEE, 2016.

Xie, Xianghui. "Low-power technologies in high- performance computer: trends and perspectives." *National Science Review* (2016): nwv087.

Xun, Yang, et al. "A platform for system-on-a-chip design prototyping." *ASIC, 2001. Proceedings. 4th International Conference on*. IEEE, 2001.

Zhang, Bin. *FPGA Design of a Multicore Neuromorphic Processing System*. Diss. University of Dayton, 2016.

Zhang, Chi, Ren Chen, and Viktor Prasanna. "High Throughput Large Scale Sorting on a CPU-FPGA Heterogeneous Platform." (2016).

Zhu, Maohua, et al. "CNNLab: a Novel Parallel Framework for Neural Networks using GPU and FPGA-a Practical Study with Trade-off Analysis." *arXiv preprint arXiv:1606.06234* (2016).

Printed in the USA
CPSIA information can be obtained
at www.ICGtesting.com
LVHW010713171023
761330LV00005B/75